TERRA II
...A WAY OUT
THE STARSEED TRANSMISSION

TERRA II

...A WAY OUT

THE STARSEED TRANSMISSION

Transmitted by

Timothy Leary

Lynn Wayne
Benner • • • • • • • • • • • • • • Guanine

Joanna Leary

CENTRAL CONTROL

REPRODUCED COMMUNITY

TERRA VALLEY

VIEW OF TERRA VALLEY

REFRIGERATION

ELEVATOR

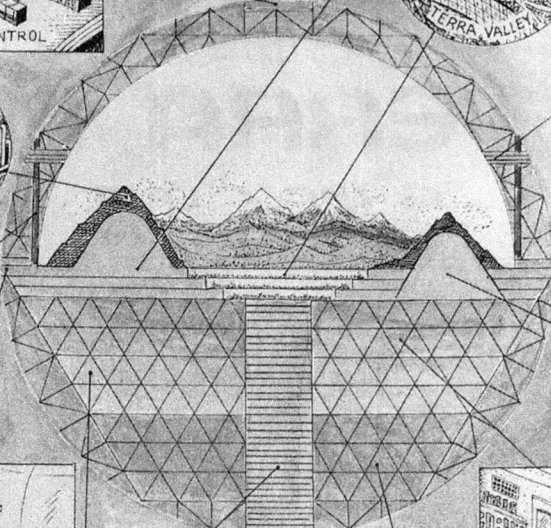

LOWER LEVEL CITY

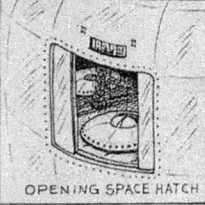

OPENING SPACE HATCH

WAREHOUSING

RAW MATERIAL

OFFICES & LABORATORIES

MANUFACTURING & ASSEMBLY PLANTS

SHUTTLE CAR AT PARK STATION

OPEN AIR THEATRE

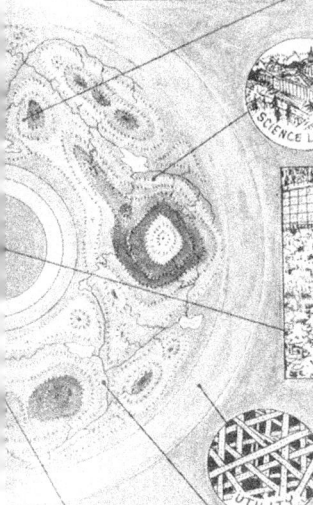

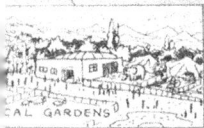

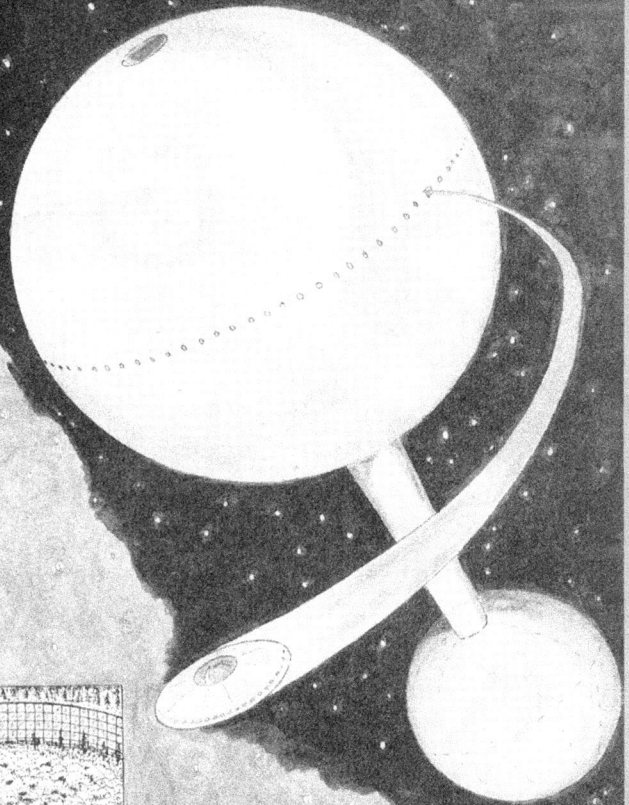

RING APARTMENTS

WALKS

SKI LODGE

SCIENCE LIBRARY

VALLEY FLOOR

UTILITY

CAL GARDENS

HOSPITAL

TERRA II •

TRANSMITTED BY

TIMOTHY LEARY

L.WAYNE BENNER

GUANINE

JOANNA LEARY

TRADE MARK DESIGN: STARSEED, INC

Ima chikywo no seimei ni totte wakuseiteki-shikyu wo deru toki.

Seimei no sushi wa yonjuokunen mo mae ni anatagata no wakuseijo ni konnichi made no zanji na shinka no aojashin wo tsutsumi shimeshita "genshikaku-choryu" no hangata ni yotte seishoku sareta.

Shinka no mokuteki wa wareware ya anatagata no hoshioya ga matsu ginga-m0 to renraku ya, soku ni kaeru kotogadekiru shinkei -s0shiki wo kaihatsu suru koto da.

Anagata wa ima "dai-nana zuno-men" e mukatte yochuki no hentai no naka ni a ri, arumono wa kaeru tabi ga dekiru butai ni iru.

Masani anatagata wa hassei no kagaku-k0z0-shiki no naka ni, towa eno kagi wo mitsuke yoto shite-ori, sono naka ni seimei no seiten wo miidaso to shiteiru. Anatagata ga fumetsu no sekinin wo tsuikosuru toki ga kiteiru, shiniyuku to iukoto wa fuhitsuyo na koto da.

Anatagata wa shinkei-soshiki no kagaku no naka ni hitotsu no kagi wo miidaso to shiteiru, Kenmei ni arushu no kagaku-k0seibutsu wo tsukaeba anatagata no shinkei-soshiki wa hassei-hosoku wo kaidoku dekiru daro.

Anatagata wakuseijo no subete no seimei-gensh0 wa hitotsu. Subete no seimei wa ima ie ni kaera nakareba naranai.

Zentai no jiyu, sekinin, shuzokukan no sogo-chowa wa tabi wo kano ni suru daro. Yochuki-teki-jotai ni aru anatagata no minzoku, bunka, kokka no ninshiki wo koenakereba naranai. Anatagata ga yaranakereba naranai tada hitotsu no koto wa "ikiru" to iu koto da. Anatagata ga ikinobiru tame no yuiitsu no hoho wa ie ni kaeru to iu koto da.

Nihon no hitobito wa anatagata no wakuseijo de mottomo susunda jinshu no hitotsu de ari, hoka no nakama nimo ooku no bensen wo ataeru daro.

Wareware wa itotsu no shingo to shite anatagata no taiyo-kei ni suisei wo okutte iru.

Anatagata no seishi-sen wa chikyu-seimei no hana da. Tojoin ga hensei sare, tabi ga hajimareba, anatagata no wakuseijo kara wa tadachi ni senso ya hinkon, soshite nikushimi ya osore wa kieteyuku daro. Soshite taiko no keiji ya tenkai no bijon ga chikaku sareru daro. Soshite anatagata ga ie ni tsuita toki atarashii oshie to chikara ga anatagata wo machiukete iru koto daro.

Zugai wo buchi yabure!

It is time for life on earth to leave the planetary womb and learn to walk through the stars.

Life was seeded on your planet billions of years ago by nucleotide templates which contained the blueprint for gradual evolution through a sequence of bio-mechanical stages.

The goal of evolution is to produce nervous systems capable of communicating with and returning to the galactic network where we, your interstellar parents, await you.

Life on the planet earth has now reached its halfway point, established itself, and evolved through larval mutations and metamorphoses to the seven brain stage.

At this time the voyage home is possible.

Assemble the most intelligent, advanced, courageous of your species, divided equally between men and women. Let every race, nationality, and religion be represented.

You are about to discover the key to immortality in the chemical structure of the genetic code, within which you will find the scripture of life. The time has come for you to accept the responsibility of immortality. It is not necessary for you to die.

You will discover the key to enhanced intelligence within the chemistry of your nervous system. Certain chemicals, used wisely, will enable your nervous system to decipher the genetic code.

All life on your planet is a unity. All life must come home.

Total freedom, responsibility, and interspecies harmony will make the voyage possible. You must transcend larval identities of race, culture, and nationality, your only allegiance is to life. The only way you will survive is to make the voyage home.

The Japanese people are the most advanced race on your planet and will give protection to the company.

We are sending a comet to your solar system as a sign that the time has come to look to the stars.

When you arrive back home you will be given new instructions and powers. Your sperm ship is the flower of terrestrial life. As soon as the company is formed and the voyage begun, war, poverty, hatred, fear will disappear from your planet and the most ancient prophecies and celestial visions will be realized.
Mutate!

Come home in glory.

Terra II ...A Way Out

Hilaritas Press ISBN Print Edition: 978-1-952746-30-7

Hilaritas Press ISBN eBook Edition: 978-1-952746-31-4

First Edition 1974, Imprinting Press
Second Edition 2024, Hilaritas Press

Cover Design by Richard Rasa
Book Design by Pelorian Digital
Artwork by fellow inmate Harold W. Olson
Some images from NASA's Hubble Telescope

Hilaritas Press, LLC.
P.O. Box 1153
Grand Junction, Colorado 81502
www.hilaritaspress.com

Acknowledgements

The Terra II voyage is feasible now because of recent advances in neurology (consciousness expansion), genetics (immortality), astronomy, and nuclear physics (propulsion).

Some of the theories and facts relating to astronomy and propulsion physics have been quoted or paraphrased from three important books! 1.) *Universe, Life, Mind* by I. S. Shklovskii and Carl Sagan, reprinted as *Intelligent Life the Universe*, Delta, N.Y. 1966; 2.) *We Are Not Alone* by Walter Sullivan, McGraw-Hill, N.Y. 1964; 3.) *Voices From the Sky*; previews of the coming space age by Arthur C. Clarke, Harper and Row, N.Y. 1965.

The authors gratefully acknowledge their indebtedness to Shklovskii, Carl Sagan and Linda Salzman Sagan, Sullivan, and Clarke, and highly recommend their books to readers interested in the technical aspects of the voyage.

The illustrations of Terra II in this edition were executed in Folsom State Prison by Harold W. Olson.

Original cover design by Francis Busco and Dana Reemes.

The first edition was limited to one thousand numbered copies, signed by the publisher.

Editor's Note 2024

Many thanks to Nick Helweg-Larsen for lending Hilaritas Press his copy of *Terra II*. Of the 1000 numbered original copies of the first edition, Nick had number 18. Here's a picture of that paragraph in his copy

THIS FIRST EDITION IS LIMITED TO ONE THOUSAND NUMBERED COPIES, SIGNED BY THE PUBLISHER, OF WHICH THIS IS NUMBER 18

We used copy #18 to ensure our new edition was as close as possible to the original. We did make some changes.

• Online retailers will not sell a book that does not have the title on the cover, so the new edition does.

• We changed the font style used in the book from what is called "Small Caps" to a more common body text font style that we think is easier to read.

• The original book had rather poor photos of space that were maybe okay for the time, but we replaced them with high resolution photos from the Hubble telescope.

• There was one graphic in the book that was poorly scanned, taken from Antoine de Saint-Exupéry's novella, *The Little Prince*. In order to avoid copyright infringement, and to present a "better" Little Prince graphic, we created an AI version that combines *The Little Prince* with the *Terra II* spacecraft. The one surviving author of the book, Wayne Benner, agreed to all of these changes.

Thanks goes out to Lynn Wayne Benner, Timothy Leary's son Zach, and Joanna Harcourt Smith's daughter Lara for their blessings and encouragement. Special thanks to Wayne Benner for his help with both the introduction to this new edition, and for his advice on the production of the book.

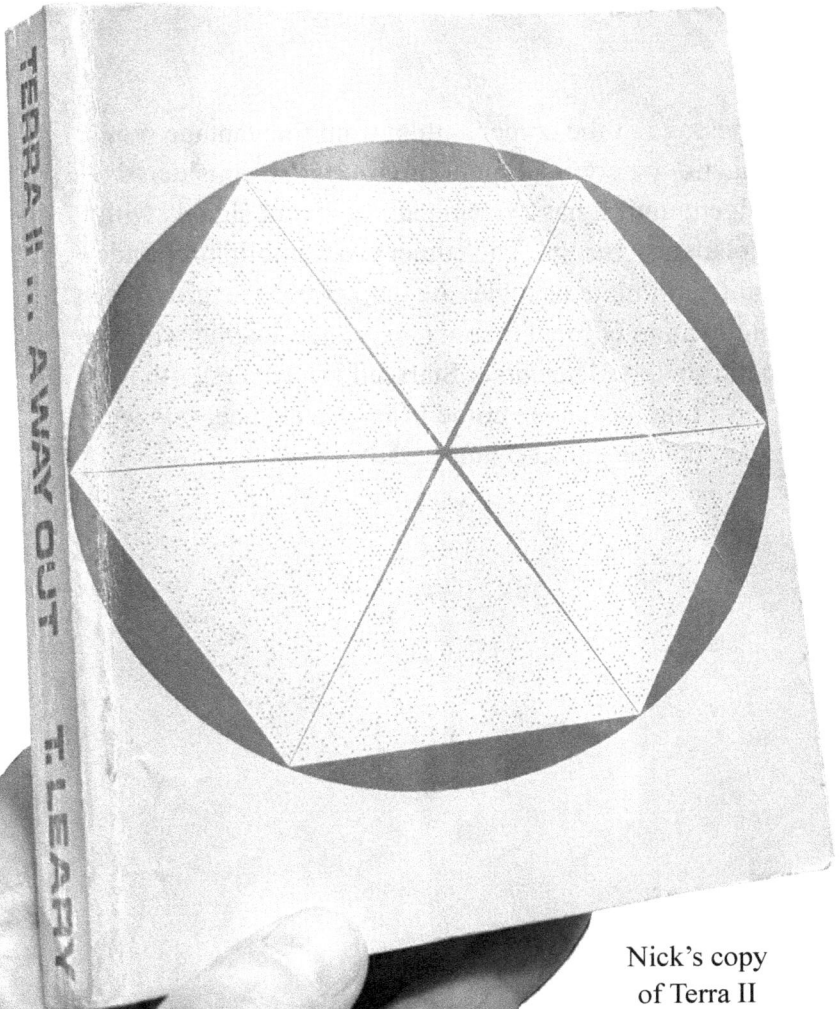

Nick's copy
of Terra II

The message of the comet is this: from the vantage point of galactic society intelligent life cannot be considered intelligent until it has exchanged intelligent signals with extraplanetary beings. The comet is a humbling reminder – a signal to stimulate us to devote our central energies to start acting in an intelligent manner – communicating with the galactic society. The comet Starseed is a reminder that we are not earthlings – that we come from outer space, that we are members of the galactic family.

"Better to be dashed to pieces as a jewel than remain perfect as a tile." is a motto dear to the Japanese soul. It refers not just to the Samurai dedication to earthly goals but to a spiritual transcendence of the technological.

TERRA II
THE STARSEED TRANSMISSION
Contents

2024 Introduction
by Oz Fritz

"Starseed is the message. *Terra II* is the realization of the message."
— Wayne Benner

"Q. 'Open the gates of your hearts and be free. Break out. Follow me to freedom, love, laughter.' Is this an indication that you were going to escape and that they should break out with you?"
— from Timothy Leary's escape trial from *Changing My Mind, Among Others* (Prentice-Hall, 1982)

"Escape is the message of my life in every form."
— ibid.

"Everything that we can think of will change."
— Benner, 2024

The overwhelming desire to escape seems the natural inclination for a caged human being. It reaches to the core of one's existence deeply affecting both the conscious and subconscious mind. If proven to be physically untenable, other ways to get out are found by the determined individual.

I spent a week in solitary confinement in 1977 for possession of marijuana due to an overcrowded prison system. The first night of incarceration I had vivid dreams of soaring far and wide in distant and expansive realms. I felt happy and free until I woke up back in my imprisoned body.

Timothy Leary spent several years in and out of multiple jails stemming from the Socratic crime of corrupting the youth with new ideas. The judge literally gave this reason when sentencing him to 30 years in jail for allegedly possessing two partial joints of weed – Leary claims the arresting officer planted it on him. Fast forward to his final bust following a dramatic jail break and nearly 2 years on the lam in Europe seeking political asylum on philosophical grounds. In transit with his partner Joanna waiting in line to go through Border Security in Afghanistan with their passports out, an agent of the U.S. government snatched their passports out of their hands and ran off. They were arrested for attempting to enter the country without proper documentation, jailed and eventually turned over to the American authorities.

Back in the United States, following his trial for escaping they threw him into the hole in Folsom Prison next to Charlie Manson. He compared it to a stellar Black Hole, a region of spacetime in the cosmos from which no gravity can escape. This is the context that framed the Starseed Transmissions and the writing of *Terra II*, a translation of said transmissions. The transmissions resulted from a series of occult experiments by a circle of four: two couples – Tim and Joanna Leary along with Wayne Benner and his partner Guanine. The men were locked up in Folsom, the woman free to support their cause on the outside. They were allowed one visit a week. Leary called this the most productive intellectual period of his life. *Neurologic*, the first writing on the eight-circuit model of consciousness, was his first essay in prison and is considered part of the Starseed material though composed before the group formed. It was written on the back of a legal brief and smuggled out of jail.

Toward the end of *Terra II*, Timothy gives some details of the experiments, but there's a sense he doesn't reveal all. The group was modelled on Renaissance-era English occultist John Dee, and his partner Edward Kelley. He presents a creatively original and scientific interpretation of what Dee and Kelley were up to.

Terra II is a spaceship meant to convey a microcosm of the Earth's eco-system and a crew of 5000 on a voyage to the center of the universe and back. Terra I is Earth. Every aspect of the project is planned in great detail. Terra II and I also double for the higher and lower brain circuits in Leary's eight circuit neurological model of the nervous system. Leary would ask Benner a question then sometimes say his response was a Terra I answer.

Terra II is an escape vessel on multiple levels. Leary and Benner escaped with their noncorporeal presence and attention by coming up with a new vision for humanity: a blueprint for how to get to the stars with the hope that contact would be made with Higher Intelligence. This intelligence wasn't automatically assumed to be from an alien race, though that wasn't ruled out. Leary read extensively on exo-biology, biology off Earth, a new science pioneered by Carl Sagan who was also very involved with SETI – the search for extra-terrestrial intelligence, another area that greatly interested Tim. In *Terra II* he plans to set-up a radio antenna on the dark side of the Moon free from the radio signals coming from Earth to increase the chances of receiving any cosmic messages. Elsewhere he speculates that this Higher Intelligence could come from humans without

saying what that would look like – ourselves in the future?
An advanced race of humans from another planet? He doesn't
speculate.

Everything that went into writing the book elevated
their awareness and mood far out of the gray monotony of
prison. The creation of a Utopian concept and its realization
– the construction and operation of a space-age Noah's
Ark the size of a small city stands in stark contrast to the
restricted environment where it began. Leary redefined the
negative of a brutally harsh, involuntary confinement by
modelling it after a Benedictine Monastery removed from the
distractions of the world. It changed into a positive – the ideal
environment to practice and apply philosophy. This kind of
alchemy, producing gold out of a base situation is one of the
great lessons to be learned from reading *Terra II*.

• • • • • •

In 1990 I visited the Fake Sufi School in Northern
California for 3 weeks along with 200 - 300 folks from
around the world for a workshop and their annual Labor
Day convention. This would be the last convention of this
kind. The school was in the process of decentralizing and
scaling back its operation, they didn't want it to get too big.
The recommendation was for people to join forces in small
groups as opposed to working alone or in large organizations.
I attended a Timothy Leary talk soon after returning to
New York and he was stressing the exact same point; it felt
like an extension of the conversation from California as if
Tim was an undercover Fake Sufi: work in small groups
toward a common aim. The group that received the Starseed
Transmissions seems a perfect example.

Joanna Harcourt-Smith first heard of Timothy Leary
through Anita Pallenberg at her paramour Keith Richards'

estate in the south of France called Villa Necôte. The Rolling Stones had recently finished recording most of their album *Exile On Main Street* in the basement there. Pallenberg and Harcourt-Smith dreamed up a longshot plan to get the Rolling Stones to donate profits from their upcoming tour to the 1972 George McGovern presidential campaign against Richard Nixon. The campaign was running out of money. "McGovern will pardon Timothy Leary when he's elected," predicted Anita. Joanna asked, "who?" Pallenberg gave a glowing report of the fugitive philosopher adding that he was attractive too. Anita knew Leary from his occasional indulgence in the party scene at Villa Necôte when the Stones were recording *Exile*. His presence there seemed appropriate since he was also an exile on Main Street having been on the run from the U.S. Justice system for over a year at that time.

Harcourt-Smith was born into wealthy, old European society. Her step-father, Árpád Plesch was a well-known financier and banker. Her grandfather, Sir Cecil Harcourt-Smith curated the Greek and Roman antiquities at the British Museum in the early XXth Century. Prince Rainier and Princess Grace of Monaco had been among the guests at her first marriage in 1966. Coincidentally, years before meeting Tim, she had dated Michel Hauchard, an older, shadowy, well-connected gangster who financed Leary's stay in Switzerland with a book deal after he fled California. Hauchard claims to have tried to persuade the Swiss government to grant Leary political asylum. She reconnected with him on her trip to America in September, 1972 to talk to the McGovern campaign. Leary's then benefactor gave her his impressions of Tim.

Her curiosity piqued, with the help of Hauchard, she tracked Leary down upon returning to Switzerland in November, 1972. They got together as a dyadic unit almost immediately. Harcourt-Smith tells the story in her excellent memoir, *Tripping the Bardo with Timothy Leary (*Future Primitive Books, 2013*)*. "Bardo" comes from the Tibetan Buddhism word for the space a person's soul enters after the body dies and before it takes rebirth into another body or achieves liberation from human existence. She framed it that way because they were together exactly 49 days before he was recaptured in Kabul; 49 days is the length of a Bardo cycle. They took LSD every day during that period. In the Preface she writes: "The Bardo was why Timothy Leary took LSD so often – not to 'party' or get high, but to learn how to 'die' and trail-blaze the final frontier of consciousness." (p. xxiii).

There's speculation that Leary knew he would eventually be recaptured and sought out a partner he could recruit to be a trusted advocate on the outside. I'm skeptical he planned it since it was Joanna that sought out Tim. It seems more like synchronicity, to me. Whatever happened, they fell deeply in love. Joanna abandoned the hedonic comfort of a young jet-setting European socialite and tirelessly took up the good Doctor's cause, not only to free him from prison, which she did after 3 ½ years, but to promote his writings and philosophy. In *Flashbacks* (Tarcher, 1983), Leary's autobiography, he writes that she was able to have her name legally changed to Joanna Leary to give the appearance they were married. She writes that she found it strange Leary gave her a co-writing credit on *Terra II* "because of the information he received through me as transmissions." She mentions working out of the Starseed office in San Francisco which doubled as the Free Tim! Headquarters and was financed by film director Francis Ford Coppola. Incidentally, I reviewed Harcourt-Smith's memoir when it came out. She loved the write-up, got my number through a mutual friend, and called me up to let me know and to encourage me to keep writing.

Another highly serendipitous synchronicity for Leary occurred when he met Wayne Benner in Folsom Prison, the second member of the Starseed core group on the prison side. It seems the best thing that could have happened to either of them. They changed each other's lives and stayed life-long friends long after they'd been freed. It was Benner who picked Leary up when he finally got released in 1976. Later, they would lecture together at Star Trek conventions. Leary gives a beautiful description of Benner in *Terra II*.

Lynn Wayne Benner is an extremely interesting person. He presents the story of his life up to his final release from prison in *Seven Shadows*, an autobiography that appears self-published.

Benner turned to robbing banks in the early '60's after getting ripped off in a business deal and suddenly finding himself unable to support his young family. He got away with it for about two years before getting caught and sentenced to life in prison after a kidnapping scheme went wrong. He had just turned 21 when he entered the penal system. Several escape attempts were made which ultimately resulted in incredibly abusive treatment by prison authorities including being sentenced to months at a time in solitary confinement on a reduced diet. Or receiving his small daily ration of water obviously contaminated with urine. Some of the prison stories are truly horrifying.

At some point he found himself next to a wise inmate who hipped him to prison and street politics. "I did feel a slight crack in the 'wall of ignorance' as he called it that had me blindly accepting the 'dogma' of authority" (*Seven Shadows*, p. 188). Later, after getting transferred to Folsom and placed once again in solitary, a prison Chaplin visited him and offered to get him a *Bible* to read. After finishing the *Bible,* the Chaplin got him a *Koran.* With the help of the Chaplin, he ended up reading everything he could acquire on religious studies and theology. That led to the Chaplin bringing him volumes of an Encyclopedia once every two weeks. He read the entire set twice. He became an inveterate reader and enthusiastic autodidact. "Books calmed my mind, soothed my spirit, eased the pain in my heart and freed my soul" (*ibid.,* p. 203). "So, I read and waited and read and waited. There was so much to learn. Each book was a delight, a chat with brilliant people throughout the ages. I soaked my mind in the pool of a million bright ideas" (*ibid.* p. 210).

After another failed escape attempt and a return to the hole, Benner was fortunate to have a lawyer visit him who let him know that he could request certain legal documents and books. This led him on a path to becoming an advocate for prisoner rights.

Benner writes that when Leary arrived in Folsom, all the gangs and factions wanted to claim him for their own to take advantage of his notoriety. A jail house lawyer named Bob Hyde asked Wayne to mediate between the rival groups to avoid conflict. Wayne advised Tim to stay neutral and not join any of them. "It was my opinion that the prison authorities sent him there to get him killed . . . The peace was made and I made a friend. . . . Timothy was the brightest person that I had ever met. Talking to Tim gave life and substance to the raw data that I had gleaned from books" (*ibid.* p. 279). Listening to a recent interview of Wayne Benner by Mike Gathers it soon becomes obvious that he is just as much of a philosopher and just as interested in changing the world as Timothy Leary. He is also an inventor who holds at least six patents, according to Leary. Wayne also had some very progressive economic ideas regarding the redistribution of wealth and finding productive uses of increased leisure time stemming from menial jobs being replaced by automation and AI. Leary liked these ideas, but didn't think they entirely fit in with the Terra II project so he suggested Wayne reach out to Robert Anton Wilson. The day he was released from prison in 1975 he was given 24 hours to get out of California. A wealthy jewel thief Wayne helped in prison arranged for a limo to take him to the airport. He changed his plans and opted to have the driver take him to Berkeley to visit a friend of his he had met through Joanna. Coincidentally, unbeknownst to Wayne, this friend lived in the same apartment building as Wilson. The friend offered to provide an introduction. Benner accepted, and ended up having a 3 hour conversation with Bob that included his economic theories. Using the Discordian pseudonym, Mordecai the Foul, Wilson wrote an essay on various proposed progressive economic theories that included Benner's. It is called "The RICH Economy" and is included in the collection, *The Illuminati Papers*. (And/Or Press, 1980)

Benner's girlfriend was the fourth member of Starseed. Not much is known about her including her full name. In *Seven Shadows*, Wayne calls her Sigrid. She was a legal reporter for a Sacramento newspaper. "Sigrid was young, tall, slim and energetic." They met when she requested an interview with him. Around the time Benner and Leary began collaborating, Sigrid said she was in love with him and wished to get married. They got engaged. Sigrid and Joanna became good friends. She is referred to as Guanine in *Terra II*; Robert Anton Wilson writes in *Cosmic Trigger* that was the only name she preferred to go by. Guanine is one of the four basic amino acids found in DNA and RNA. Joanna was Adenine. "The model for Terra II is DNA. There will be many spiritual models and mythic visions which could guide and pattern such an expedition."

Chapter 19 describes how the Starseed Transmissions are based on "images, revelations, visions received by a quadro-magnetic human unit." It's tempting to consider these communications originating from an extraterrestrial Intelligence in Outer Space. It seems equally possible that Starseed refers to DNA as it suggests the theory of panspermia which holds that the building blocks of life arrived on Earth from a meteorite. Thus, the transmissions could have originated from deep inside, from communication with their DNA. Panspermia was one of the first ideas Benner and Leary discussed and is key to Starseed.

The resonance between Leary and Benner with the English polymath John Dee and his assistant Edward Kelley is striking. Dee was one of the most learned men of his time with feet firmly planted in both science and mysticism. As a mathematician and astrologer, with the exception of Mercury, he was able to map out the positions of the known heavenly bodies in his birth chart with a great deal of accuracy. Like Leary, he got busted for his activities – charged with treason for casting a horoscope for Mary, Queen of Scots, but was able to clear his name. Later, his magical work got him accused of consorting with demons despite a strong foundation in Christian belief. In his early twenties he travelled abroad seeking further education and struck up a close friendship with the influential mapmaker Gerard Mercator. Only a few years after Copernicus' radical theory that the Sun, not the Earth was the center of the universe became known, Dee and Mercator were busy experimenting with new models of the universe. His knowledge of mapmaking later led to him becoming an advisor to the explorer Martin Frobisher who set off in search of the fabled

Northwest passage in hopes of expanding trade with Asia.

Dee became a passionate collector of books and rare manuscripts. He tried to get the English monarchy to establish a national library, but this never got off the ground. His own extensive library attracted many scholars becoming a highly respected source of knowledge. One such occult manuscript was called *Steganographia* written by one of the founders of modern cryptography, the German Benedictine abbot Johannes Trithemius in the early 16th Century. When Dee heard of an extant copy, he travelled to Europe and painstakingly copied it out by hand. The cryptography in this work was closely related to Cabala. He thought that these techniques could be used to decipher other obscure texts.

Dee rose to prominence as the astrologer, advisor and confidant to Queen Elizabeth I who had a strong interest in magic. "Elizabeth had a strong sense of the forces of the cosmos acting upon her, and regarded her monarchial powers as in some way magical." (Wooley, *The Queen's Conjurer*, Henry Holt and Co., 2001). He chose her coronation date based on the stars. Some sixty plus years after his death, the oldest inhabitant of Dee's village described him as '"very handsome', tall and slender with a fair complexion." She "recalled how children ran screaming from his door because he was 'accounted a conjuror' and how he would act as a 'great Peacemaker' among squabbling adults." (*ibid*, p.81).

Edward Kelley initially showed up at Dee's residence under an assumed name with the intention of hiding out from the law. He was a reputed con man and swindler rumored to have been convicted of forgery, but "was unusually well educated for a man of his status" (*ibid*, p. 131). He also turned out to be an extremely talented "skryer," someone able bring back communications from nonhuman entities by staring into a crystal stone. Their first major contact was with the Archangel Uriel followed soon after by Michael. Their most famous occult reception was the Angelic or Enochian language.

It's not clear what role, if any, Dee and Kelley's wives played in their occult experiments. It's known that Kelley claimed an Angel told Dee they should sleep with each other's spouses causing the magician great consternation. A rift between Kelley and Dee developed not long after.

I suspect the women of the Starseed group had a much stronger and more essential role. Leary describes them as charging the workings with solar love with their weekly visits to the two prisoners. Benner gives a hint of how this may have helped his skrying in *Seven Shadows* (p. 280): "Love is such a powerful energy. My entire being glowed. My mind was aflame with ideas and thoughts. My heart radiated intense joy."

Terra II overflows with the optimism of the 1960s and early 70s, but it's no pie in the sky fantasy. The practical aspects of how to get the project off the ground and bring it to fruition have been worked out in great detail. It includes an estimation of how much it would cost, proposals for its financing, where, when and how initial planning would begin to a rough outline of the architecture of the craft as well as a consideration of the problems and dynamics of having a crew of 5000 live together for an extended period of time. However, it does lean in the direction of science fiction though jumping off from the platform of science and speculative science.

Leary assembles a formidable team of scientists and science fiction writers to legitimatize and reinforce the Starseed vision quoting some of them at great length. These include Carl Sagan, I.S. Shklovski, Arthur C. Clarke, Francis Crick, Leslie Orgel, Robert Heinlein, Walter Sullivan, Sir

Bernard Lovell, Eugen Sänger and others. He hopes to get John Lilly and Jean-Paul Sartre on board in the planning stages. There's reference to experiments carried out in the Soviet Union by the Krasnoyarsk Institute of Physics of keeping humans in a closed biological environment for an extended time for the purpose of working out the problems for long-term space travel. Solving those problems first on Earth is seen as a necessary step for making Terra II work.

The Biosphere 2 project, starting in the late '80s, repeated this experiment. From 1991 to '93 and again from March to September of '94 eight people lived in a completely sealed biological world built near Oracle, Arizona that unfortunately ran into some serious problems with both the humans and the technology. In the long term it aimed to provide data for closed eco-systems necessary for space colonization. It was the brainchild of systems ecologist, engineer, theater producer, and writer John Allen aka Johnny Dolphin and his group/school The Theater of All Possibilities. It was financed by billionaire philanthropist Ed Bass. It seems likely that Dolphin got inspired by Leary's ideas for space migration in *Terra II.* Buckminster Fuller's *Operation Manual for Spaceship Earth* was another seminal influence. Bass, Dolphin and Kathelin Hoffman had opened a performing arts center in Fort Worth, Texas called the Caravan of Dreams where Leary and other major players in the avant garde would perform; people like Ornette Colemen, William Burroughs, Brion Gysin, Sun Ra, etc. They also had a record label releasing albums by Coleman, Ronald Shannon Jackson and James Blood Ulmer.

• • • • • •

One of Timothy Leary's many talents was coming up with succinct phrases or slogans to encapsulate philosophical ideas. At the height of his psychedelic guru persona in early

1967 at the Human Be-In in San Francisco's Golden Gate Park he introduced the motto "turn on, tune in, and drop out" to a crowd of some 30,000 hippies, seekers and political revolutionaries. The saying became forever entrenched in the cultural psyche and, despite the initial context, applies to more than simply taking psychedelic drugs. Another of his pithy catchphrases: "you can become anyone this time around" is the title of a spoken word album he made in 1970 to raise funds for his California gubernatorial campaign. It featured musical contributions from Jimi Hendrix, John Sebastian, and Stephen Stills and was recorded not long before his legal problems became a serious hindrance. In Folsom Prison circa 1973 a new philosophical credo was about to take shape.

Leary's optimism was no more evident than the promotion of the SMI²LE formula which stands for Space Migration, Intelligence Increase and Life Extension. Wayne Benner says it was the first result of their prison collaboration prior to *Terra II*. The entire book can be seen as one long explication of those principles. Space Migration is the purpose of Terra II. Chapters 14 and 15 are devoted to longevity and immortality. Leary writes that intelligence increase is absolutely necessary for life extension. He credits Benner with coming up and combining these three prerogatives for human evolution. Tim probably came up with the acronym.

SMI²LE has a featured role in *Exo-Psychology* (Leary, Starseed/Peace Press, 1977) an important book for explaining the eight-circuit model of the human brain and nervous system. It expanded upon *Neurologic*. It starts: "The purpose of 'Life' is SMI²LE. " The next page has a minimalist poem similar in style to his translation of the *Tao Te Ching, (Way of Life)* that became *Psychedelic Prayers*. The poem reaches the conclusion to SMI²LE. The Preface elucidates the SMI²LE objectives presenting the evolutionary philosophy first outlined in *Terra II*.

Chapter 13, "An Interstellar Neurogenetic Teleology" once again focuses on SMI²LE. Tim repeats the opening dedication on the aim of life then writes: "We are designed to Use our Heads (I^2) in order to Use Time (LE) in order to Use Space (SM). Of these three associated imperatives Intelligence Increase is the most important. When the human has learned how to use the brain to:

4. selectively re-imprint the four terrestrial circuits

5. control the body

6. master the creation of multiple realities by means of serial re-imprinting

7. imprint (i.e., experientially identify with the DNA code)

8. decipher nuclear-quantum intelligence

Then Life Extension and Space Migration will be attained." The numeration reflects the task of the different circuits starting with C4, the summation of the four lower circuits necessary to function on Terra I. I guess we haven't become smart enough yet.

Prior to Benner, The Beatles may have inspired Timothy to SMI²LE. They opened their landmark 1967 album *Sgt Pepper's Lonely Hearts Club Band* singing:

"It was twenty years ago today Sgt. Pepper taught the band to play.

They've been going in and out of style, but their guaranteed to raise a *smile*."

Leary first heard the album when John Perry Barlow and Bob Weir, lyricist and rhythm guitarist for the Grateful

Dead, brought it up to his then residence in Millbrook, NY. After the record finished, Tim dramatically proclaimed "My work is finished. Now it's out." He didn't mean his life work, but his role as a psychedelic guru. Later he remarked on *Sgt. Pepper*, "The newest message from Liverpool is the Newest Testament chanted by Four Evangelist-saints, John, Paul, George and Ringo." In *Timothy Leary, A Biography,* Robert Greenfield writes that Tim and his wife Rosemary spent an hour or two each day listening to *Sgt. Pepper* and *Their Satanic Majesties Request* by the Rolling Stones. About the former, Tim remarked, "I've got nothing to say that you haven't said briefer, cleaner, stronger." A couple of years later John Lennon and Yoko Ono invited the Leary's to their bed-in for peace protest in Montreal to sing in the chorus of *Give Peace A Chance.* Before departing, Lennon asked if there was anything he could do to help candidate Leary's campaign for Governor of California. Tim requested he write a song around the theme of "come together." Lennon wrote the song, it opens The Beatles last album, *Abbey Road,* but by then Leary had been blocked from the race by the California legal system. Wayne Benner remarked that all the Beatles liked Tim. "I went to all his birthday parties and there was always at least one Beatle in attendance."

Sgt. *Pepper* also unleashed the visage of Aleister Crowley into popular culture which subsequently made his teachings far more known to the public than when he was alive. He was one of The Beatles influences they had on the cover along with 57 others. Internet speculation combined with strange coincidences suggests that Sgt. Pepper *is* Aleister Crowley. Though both Lennon and McCartney knew of Crowley, I'm skeptical that was the intention. Lon Milo Duquette pointed out that Aleister Crowley died 20 years before The Beatles opened the album proclaiming, "it was 20 years ago today,

Sgt.Pepper taught the band to play . . . " What he taught the band to play is presented in his system somewhat archaically as the Knowledge and Conversation of the Holy Guardian Angel. This involves activating and opening the heart to a strength and degree largely unknown, much less realized in contemporary society. The club consists of "lonely hearts" because not many have attained. "Pepper" symbolizes this activation in its correspondence with the element Fire. Set the heart on fire. This trope appears both explicitly and implicitly in *Terra II* as we shall see.

In a letter to Robert Anton Wilson, Leary remarked upon the extraordinary synchronicities he had with Crowley. He had a very powerful mystical experience in the same location in the Algerian desert where Crowley called forth the demon Choronzon which may have had a lasting psychological impact on Crowley's assistant in the operation, Victor Neuberg. The operation in question, told in *The Vision and the Voice* (Weiser, 1998) by Crowley, Neuberg, and Mary Desti, originated from the work of John Dee. Dee's Enochian Magick was a major source for some of Crowley's experiments. Crowley himself stated he was a reincarnation of Dee's skryer, Edward Kelley. There's a YouTube clip of Leary saying he felt he was carrying on Crowley's work. The Starseed philosophy closely aligns with Crowley's system, Thelema, which means "love under will." The word itself is Greek for Will. By Greek Cabala, Thelema adds to 93 as does the word Agape, divine love. There is more than one sentence in *Terra II* of this nature: "We believe love will energize our voyage, love will guide and love will be the outcome."

The full title of the book, *Terra II: A Way Out* aligns with the magical formula Get OUT which Crowley contrived and presented in Chapter 23 from *The Book of Lies* (Weiser, 1981).

The Hermetic principle, "that which is above is like that which is below, that which is below is like that which is above", a bedrock axiom underlying all modern occultism, runs throughout the book. Some examples include comparing the four members of the Starseed group with the four amino acids found in the DNA code, the Van Allen belt protecting the Earth from radiation seen as isomorphic to the embryonic membranes protecting the human fetus, and the suggestion to think of the ship as a human body, the crew as the nervous system and the individual crew members as neurons. The axiom itself is quoted in the abbreviated form, "What is above, so below." Leary affirms this principle with Science: "[s]cientists tell us that the same laws of energy transformation occur at all levels of the cosmos. The same processes which govern the movement of nuclear particles direct the movement of galaxies." On another mystical front in the 8th transmission, the space-ship Terra II gets referred to as a blossom on the kabbalistic Tree of Life.

Terra II formulates a large-scale act of Magick, causing change to occur in conformity with Will. It's meant to bring the entire planet together in a common aim thus ending war and conflict. Leary sees himself as an alchemist: "[t] he alchemist and his mate wait, sometimes for thirty years, disclosing substances, boiling elements, breathing fumes, and blowing on the flames Praying. And then, when the elements are in order, the stars are in position, the heart's love pure: it happens." The discovery of a new comet by Luboš Kohoutek in 1973 streaking toward the solar system was seen by the group as a cosmic affirmation. It generated a lot of publicity because it was predicted to be very bright in the night sky. However, it didn't live up to its luminous expectations and was only briefly and barely visible to the naked eye in January, 1974.

By their own admission, the success of the project counted on great advances in engineering and technology in the next twenty-five years. A moon base would need to be built and established where construction of the space-ship would take place. Leary also refers to Terra II as a star-ship and in other instances, a time-ship. The latter moniker given due to the effects of time dilation predicted by Einstein's Special Theory of Relativity which states that the passage of time depends upon the frame of reference of the observer. A passenger on the time-ship hurtling toward the center of the universe might only experience a 20 – 30 years of flight time, but find that thousands, or millions or billions of years may have gone by on Earth, depending on whose calculation proves correct – there are at least two different predictions on the extent of the time dilation given in *Terra II*.

They admit that means of the propulsion to power the ship were unknown. Chapter 20 presents plausible sounding scientific speculations on what that propulsion might look like. At least one of their proposed technological breakthroughs did come to pass, for better or for worse. Computers were envisioned to play a significant role. What sounds like personal computers, unheard of at the time of writing, were included in the plans. Three years later, Steve Jobs and Steve Wozniak introduced the first Apple micro-computer to begin the home computing revolution. The computers on Terra II were planned to have the ability to talk to one another some ten years before new communication protocols established the modern internet and more than twenty years before the World Wide Web became ubiquitous.

• • • • • •

Some time ago I saw a clip of a prisoner Leary interview

from that era where he said with a big smile on his face, "yeah, we tried to change the world. We knew the odds were against us, but we had a sense of humor about it." Every crew member on Terra II is to learn how to effectively operate and direct their nervous system via "continual experimentation with a sense of humor." In other words, they need to work on themselves in some way – yoga, martial arts, meditation, etc., whatever their preference. But always with a sense of humor, "nothing to get hung about ..." Tim has an interesting dedication to the character most associated with the phrase, "work on self", G. I. Gurdjieff. in *Info–Psychology* (Falcon Press, 1987):

> "This book celebrates all
> Evolutionary Agents
> And Cyber–punks
> On This Planet and Elsewhere
> Especially
> Georges I. Gurdjieff
>   who reminded us to laugh
> Aleister Crowley
>   Who did the English translation . . ."

When fellow stand-up philosopher Robert Anton Wilson visited Tim for the last time, he told him that he was the funniest person he knew and he knew George Carlin. To me, that compliment is more prestigious and meaningful that being knighted by a King or Queen or receiving the Presidential Medal of Freedom. Humor seems key to the optimism Leary radiated as a real force. This force is discussed at the end of Wilson's *Cosmic Trigger Vol. I* (Hilaritas, 2016). "And so I learned the final secret of the Illuminati."

At the time of this writing the objectives presented in *Terra II* appear a long way off. They certainly didn't come close to being realized in the timeline given. Yet, slowly but surely advances are being made. There are over a dozen private companies engaged in various kinds of space exploration research and design. Space tourism is in the process of becoming available for the wealthy. Experts predict that it will be affordable and that there will be regular commercial space travel within the next 10 – 20 years. Progress may be slow and plodding but it hasn't died.

Research continues in full force on aging and longevity. In March of 2024, Dr. Eric Verdin, President and CEO of the Buck Institute for Research on Aging gave a lecture briefly outlining some of the breakthroughs in the field. He told everyone in the auditorium that they could reasonably expect to live to 90 – 95 in good health by applying anti-aging factors. He claims that we can change the rate of aging since studies have shown that 93% of longevity is determined by lifestyle with only 7% determined by genetics. The prescriptions are exactly what you would expect: proper nutrition, physical activity – he calls this the greatest anti-aging medicine we have today – and proper sleep along with stress management. Verdin says that human connection is the strongest predictor of life expectancy as well as having a sense of purpose in life. Anti-aging drugs, currently getting good results in mice, are on the horizon for humans in the next 5 – 10 years. Medicine is being re-invented to take a more proactive approach against aging. In 2021, at age 90, actor William Shatner became the oldest person to travel into Space. Shatner, of course, is famous for his role of Captain James T. Kirk of the U.S.S. Enterprise in the original *Star Trek* television series, a show that planted the idea and excitement of space exploration in the public imagination.

Some observers of historical patterns predict that civilization is entering another Dark Ages. Recent world events in the past 25 years or so don't contradict this view. In America alone we've seen the terrorism of 9/11 lead to multiple prolonged wars. Violence breeds violence and fanaticism; worldwide we've seen the growth and proliferation of new terrorist groups. A global pandemic impacted billions of people. The first major land war in Europe since W.W.II still continues to rage. The effects of climate change are no longer theoretical. The rise in political power of extreme right wing views embracing fascism seems the strongest it's ever been since the 1930s and '40s. Always a powder keg, the Middle East appears close to a major war between Israel and Iran as of this writing. The former has been at war for the past ten months with no clear indications of a resolution.

The good news is that a beacon of optimism, hope, intelligence and light continues to shine below the surface of world events as evidenced by the book you hold in your hands. The slow, incremental progression of the SMI^{2}LE vision in no way diminishes the value of this important document. Whether or not Terra II ever comes to pass, there's a wealth of practical information inside useful to any and every individual right now. It will prove inspiring to future generations. The guiding metaphor is the story of Noah's Ark – preserving life and intelligence until the flood of conflict abates. By its own admission, Leary and Benner's vision for a sane future requires "a metamorphosis, a neurological mutation."

Terra II was formulated as a response to world problems. The details of the problems may have changed, but as a species we continue to find ourselves in an emergency situation. These words from *Terra II*, written in 1973 from an off-planet perspective, seem equally true today: "Out here, beyond good and evil, one sees the human race in pain, injured nervous systems propelling robot-bodies in repetitious, aimless motion along paths labeled right and wrong." What's an Evolutionary Agent confined to Earth's gravity well to do? Leary answers with a parable he heard from his teacher in India, Shri Krishna Prem. "Have courage and keep moving."

Oz Fritz

August 5, 2024

Nevada City, CA

STARSEED —
A PSI-PHY COMET TALE

This signal is being transmitted from the Black Hole of American society. A Black Hole is a dense space with a heavy gravitational pull. Matter which falls into a Black Hole fades from view and disintegrates in the stress of gravity. Given a sufficient time, its radiation becomes too feeble to be detected from without. Although the matter of the Black Hole cannot re-escape as matter, some of it may manage to escape in the form of feeble red radiation. Some cosmologists suggest that Black Holes are the link to another realization of matter. They may be passageways to another universe, just as the manholes of Paris lead to a world beneath the street. Well, the Black Hole is a fine place from which to scan the universe. It's beyond pure undiluted bad. As good as good can be.

Out here, beyond good and evil, one sees the human race in pain, injured nervous systems propelling robot-bodies in repetitive, aimless motion along paths labeled right and wrong.

Sri Krishna Prem, the wisest man in India, sat on the floor of his little mountain-top ashram showing us pictures from medieval alchemical books. He pointed to the design of a man standing naked with devil on one shoulder, angel on the other. He said, "When you understand that, you can go on to the next lesson."

The next lesson was a parable about a great castle that was separated from shore by a swamp. Pilgrims, searchers, warriors, seeking the castle, disappeared into the marsh because each rock they stepped on sank from view. The Hero and His mate sat on the bank and watched for days. Then He rose and held His hand to her. He whispered the instructions Sri Krishna Prem transmitted to me: "Leap from rock to rock more swiftly than they sink. The trick is simple. Have courage and keep moving."

Each spot we stand on crumbles beneath us, becoming a launching pad for the next stride.

From the far future, we transmit these messages back to planet earth. Beware of the Hindu trap. It's anti-sexual. The guru, God, and the swami universe is a soft, sweet custard mush. Undifferentiated unity. True unity is contacted through increasing precision of differentiation. Psi-Phy. Philosophy of science. The universe is not chaos ruled by casual chance. The second law of thermodynamics is pessimistic nineteenth century fraud. Tensile structures emerge and lawfully evolve because of the underlying magnetic charging. Positive-negative. WoMan.

Life is an interstellar communication network. Life is disseminated through the galaxies in the form of nucleotide templates. These "seeds" land on planets, are activated by solar radiation, and evolve nervous systems. The bodies which house and transport nervous systems and the reproductive seeds are constructed in response to the atmospheric and gravitational characteristics of the host planet, the crumbling rock upon which we momentarily rest.

Evolution is concerned with nervous systems and the

sexual attractive efficiency of bodies, the preservation of future genetic codes, the expansion of consciousness.

The human being is the robot carrier of a large brain, conscious of being conscious. A robot designed to discover the circuitry which programs its behavior. The nervous system is the instrument of consciousness. When mankind discovered the function and infinite capacities of the nervous system, a mutation took place. The metamorphosis from larval earth-life to a higher destiny. The person who has made this discovery becomes a time-traveler. A Psi~Phy entity. When astronaut Mitchell saw the green jewel of earth against the black velvet expanse of interstellar distance, he became Psi-Phy. Ecology is a low level distraction. Psi-Phy boy scouts picking up trash. The genetic goal is communication. Telepathy. Electronic sexuality. Reception and transmission of electromagnetic waves. The erotics of resonance. The entire universe is gently, rhythmically, joyously vibrating. Cosmic intercourse.

This is a message of hope and interstellar love from the Black Hole. Irrepressible optimism. Yes, it is true that repressive pessimists now control planetary politics. This is an embryonic phase. Life has been evolving for 3 1/2 billion years and has just reached the half-way point.

This message of neurological resonance can be censored, imprisoned, but cannot be crushed because it comes from within, from the DNA nucleus inside each cell, from the evolving nervous system. The Higher Intelligence has already stepped on planet earth and its script is writ within our bodies, emerging in every generation.

We were wondering where to go next. Before the Japanese transmission the solution seemed to be a boat. A new society of time-ships, sailing the high seas. The Noah myth. Premonitory preparation for the emigration from planet earth. Load the boat with transmitting equipment. Radio Free Earth.

Imagine looking at a map of earth. Scattered around the oceans are a hundred little golden spots, clustering in the Caribbean, moving through the Indian ocean, the south pacific. Time-ships carrying the first global citizens. A transient community of mutants who form a new nation, who transcend their former nationalities. How do they support themselves? They are the wisest, strongest, best endowed. Mutants have to be.

Now is the time in the Sci Fi books for the cosmic intelligence agency to send the extraplanetary rescue ship.

Lovell's book on outer space presented a clue, the last chapter presents a drawing of the remnant of a living organism found on a meteorite, a nucleic acid molecule, the first signal from extraterrestrial life, help is on the way. We began to etch the design on silver pins and leatherwork. We called it STARSEED , the symbol of Psi-Phy,

In early July the New York Times carried a story about a newly discovered comet entering the solar system. Unexpected. Named after its finder, an East European astronomer, it would be visible during the fall of 1973 with a

brilliance greater than the full moon,

Here was the greatest astronomical event in recorded history appearing right on schedule, we named it Starseed , new light, new life, bright reminder of our extraterrestrial origin and future. Symbol of freedom. Guanine began telling the story on radio and TV and to everyone she met, Paul Kantner and Grace Slick offered to write a song heralding the new coming.

As the weeks went by a curious fact emerged. There was no more publicity about Starseed . Nothing in the scientific magazines. No one else had read about it. We wondered if we had dreamed it out of longing anticipation. Mysterious. Why so little publicity about the greatest light in the sky? Many mysterious things are happening these days. Watergate. The United States sinking in the swamp of history. The dollar collapses. Food shortages. Energy crisis. The worst thing to observe was the mental depression, the cerebral pollution. No one was speaking clearly about what was happening. No new solutions. A glum stagnation. A damp nostalgia. A frivolous melancholia. The Black Hole provides a clear perspective. It's up to us to provide the vision. It always happens this way. New light from the Black Hole.

And nobody said anything about the comet.

Then Paul Kantner sent in a report.

On March 1, 1973, Dr. Lubos Kohoutek, a Czechoslovakian astronomer, at Hamburg Observatory, Bergedorf, West Germany, discovered a new comet. As with most modern discoveries this one was made photographically. Later, a pre-discovery image plus numerous subsequent ones have enabled astronomers to compute the comet's orbit with considerable precision.

During July through-September, it will be too close to the

sun's direction for optical observation, but in mid-October, at a distance of about 168 million miles from the sun, its brightness should increase to a magnitude of between 8 and 12. From then on it should brighten quickly and is expected to reach naked eye visibility about mid-November. It will then be in the morning sky, in the southeast, about two hours before the sun (from San Francisco).

It is not possible to predict with precision what form the comet's tail will take nor what brightness it will attain, but indications are that it will exceed that of Halley's comet, last seen in 1910 and not due again until 1986. There is a possibility that its magnitude at perihelion (closest to the sun – only about 13 million miles) may approach that of the full moon, making it one of the brightest ever seen by man.

This confirmed the coming but renewed the question: why the silence, the lack of interest? The newspapers were filled with stories anticipating the advent of the football season. Another sign of the times. The philosophic perspective has been lost. There exists a repression, a taboo about facing the implications of the recent scientific findings which compels a total revision of our concepts of life and of human nature. Einstein's equations. Nuclear energy, the revelation of DNA code as literally a code to be deciphered. Neurological imprinting. Antimatter. Mankind clings to the old myths, avoiding the new truths.

It happened before.

"Towards the end of the sixteenth century, Giordano Bruno aroused the groggy world, asking it to fling its mind far beyond the planets. He speculated that the cosmos extended to infinity . . .

"This in itself was not so shocking; but Bruno went considerably further – he postulated a multiplicity of worlds: suns and planets with life, unseen companions for the race of man. He toyed with man's conception of himself; for this, and for magical claims and political entanglements, he was burned in 1600 . . .

"Shortly before Bruno's death, in 1600, Tycho Brahe made the first announcement of a 'new' star in the sky. A few years later he observed a comet, and proved that it moved among the planets; thus he shattered the stars about the heavens." – Charles Whitney, *The Discovery of Our Galaxy*.

Tycho's star set off excited controversy because it forced a change in the cosmology. Current theories held that the stars were fixed. But the new evidence was there flashing in the sky. The stars moved. Cosmology is not a peripheral hobby, specialty of scientific experts. Every aspect of human life is based on the answers to the cosmological questions: where did we come from? Where are we going? Tycho's star appeared exactly when Christendom was unsettled by the reformation. Luther's challenge to the immovability of catholic theology, you could get busted those days, as Galileo discovered, for advocating the idea that the earth moved.

Guanine and adenine began asking about Higher Intelligence. Guanine and Adenine are stars. They suddenly appear in the sky brilliantly transmitting radiation. Super-novae. All-out energy, high fidelity. They turn their electron-telescopic girl eyes on you searching for the signal. They were asking people they met: what's your cosmology? Never mind your sexual need, your bread problem, your ideas on Watergate. What's your cosmology? How did you get here? The seven days of genesis? The chance play of amino acids heaped up aimlessly like bricks? Does god play dice with the cosmos? If not, what's the master plan?

What's your cosmology? One young man smiled and said he was a graduate student in astronomy at Stanford. No one, he said, was certain about the path of the comet Starseed spinning into our solar system. It might come dangerously close to planet earth. Might even collide.

We began logically computing the possibilities, while, in the sky above, Sky Lab circled the globe, telescopes trained on the sun. Why hadn't Sky Lab mentioned this flying object hurtling into view, until now?

There are two alternatives: 1. The comet means danger to earth; or, 2. It will pass through our system with spectacular inoffensiveness.

Alternate one. It may be dangerous. It may disrupt the earth's atmosphere. It may smash into the earth's surface. If this were known, would it be announced by the men who control America and Russia? The Soviets have chosen Mars as their planet of choice. If the comet were to hit earth? Enormous tidal waves? Destruction of civilization? Who would survive? A weird sci-fi Noah's ark horror story. Who would end up in the safety caverns dug deep, by the joint chiefs, into the western mountains? The grim significance of the cold war takes on another dimension. We knew about it all along but it was never talked about. The very men who bomb Cambodia, provoke the Russians to accelerate the missile race, are the ones who have designed and built the secret hide-away caverns. How many cases of whiskey and tons of steak are deep-frozen in the bomb-shelter cities? Who gets to get saved? The president and his family? The military, of course. Is Agnew on the friend or the enemy list? Does Bebe Rebozo get saved? Does Ellsberg?

We recall the irritation of the Air Force with U.F.O. Reports. The negative finding of the Condon report that no extraterrestrial sightings had been confirmed, was not surprising. What disturbed was the obvious emotional bias, the fact, and this the crucial experimental datum, that the Air Force didn't want people thinking about extraterrestrial intervention. Just as the Catholic hierarchy and its Scholastic philosophers four hundred years ago didn't want people thinking that the stars might move. The ancient, basic cosmological fears and hopes. If you star speculating about Higher Intelligence visiting planet earth, a galaxy of embarrassing issues gets raised. What would the celestial visitors think of how we are running the planet? Whose selfish securities and biased superiorities would be threatened?

The Air Force U.F.O. study included, indeed, emphasized a factor which infuriated the "flying saucer" partisans. A team of psychologists studied the personalities of those who reported the sightings. How clever of the Air Force to suggest that those whose cosmologies, however vague, included the possibility of extraterrestrial intelligence, were themselves "Kooks." In wider perspective we can only endorse the Air Force psycho-diagnostic attempt. It may be that the contact with extraplanetary intelligence, the discovery of the master plan will not come via radio telescopic contact. And certainly the anticipation of "saucers" transporting humanoid bodies is naive. It is more likely that extraplanetary contact will be received by the instrument which was designed over 3 1/2 billion years ago to pick up electromagnetic vibrations. The human nervous system itself. The air force psychiatrists might have done better if, instead of administering Rorschach personality tests, they had performed intensive neurological examinations, brain-wave studies on the wild-eyed "saucer-sighters." Maybe some of the kooks carry nervous systems more receptive to electromagnetic impulses.

It is agreed that we would send out a message posing the Starseed questions. Could there be a secret conspiracy to censor extraplanetary contact? Thoughts of Dallas, Sirhan, Martin Luther King, My Lai, Cambodia, Liddy, Hunt,

Haldeman, and Ehrlichman run through the computers.

Adenine said, "If there's only four months before holocaust everyone should be told. Stop everything and finally learn about full-time love. I'm not afraid of dying. I know we'll be together. But I want to be on the same side of the wall with you . . .

The second alternative: comet Starseed flashes into our view and leaves. Each person who reads these words will, in the coming months, stand on earth looking up at this spectacular sight. Will it leave us transfigured? Lift our eyes up to universal perspective?

The comet Kohoutek, Starseed , can mean nothing or it can mean everything. It can be a reminder that this planet is just a brief crumbling stepping-stone in the voyage of life across the galaxies. That the Higher Intelligence has already established itself on earth, writ its testament within our cells, decipherable by our nervous systems. That it's about time to mutate. Get born. Create and transmit the new philosophy.

Behold a great light appears in the sky. The offer is made. The signal is flashed. Resonate with it or die eye-ground and bored.

Cosmologist Hoyle now suggests that creation emerges, not uniformly throughout the universe, but in regions of high density and intense activity, such as a developing Black Hole. Contractions and expansions take place at scattered points within the universe. Astronomers used to think that the radiation which energizes the universe originated from one Big Bang. Now it seems possible to explain creation as the product of Black Holes, whose existence has just been proven by astronomers at the University of California.

The old menopausal-stoic view of galaxies as quiescent swirls of stars, gas and dust is giving way; signs of creative upheaval are everywhere, stepping-stones in the pool of time popping up and disappearing. When Black Holes find their places in cosmology (and local politics), neurologicians will have written the most complete version of the creation story since the days of the first ancients.

The alchemy of power takes time. Neurological politics. The wizard does as little as possible. The organization is already here. It just takes the slightest move at exactly the right time to turn it on. Connect the wires. The alchemist and his mate wait, sometimes for thirty years, disclosing substances, boiling elements, breathing fumes, and blowing on the flames Praying. And then, when the elements are in order, the stars are in position, the heart's love pure: it happens. Transfiguration. The comet Starseed is the signal. The Japanese transmission is the message.

All the signs whispered it. The time had to come. Strange how everyone feels it, the dissolution of the old structure, but no one can get the perspective to see it. Americans are too close to read the portents. Superstitious is good. It means to stand above. Back away from it, climb above it for a moment and see it as a Shakespearean epic or a Greek tragedy.

There is this throne of ultimate power. The lethal crown of world empire. The curse. Roosevelt dies. Truman retires in disgrace. Ike immobilized into grinning idiocy by a heart attack. Kennedy killed. L.B.J. Ruined. Bobby slain. Nixon and Agnew revealed as criminals. Form focuses energy. It is the institution, the two-hundred-year-old structure, that is wrong. The horse-and-buggy American political system, pre-technological design, can't handle the energies released in the nineteenth and twentieth centuries. The White House should be museumed; and replaced by a towering glass pyramid, visible, shining, hooked up by two-way electronics with every neighborhood in the land.

It is time for prophecy. The omens are obvious. The moment of spiritual reckoning approaches. Karmic plague sweeps the globe. Scan the headlines. Drought. Famine. Shortage. Pollution. Malaise. Disorder. Tyranny. Espionage. Watergate is the American word for a world-wide epidemic of government illegalities. Torture in Greece. The new repression in Russia. Israeli air piracy. Libyan mania. Every week another country captured by its own military police. The Higher Intelligence scanning these developments from the high perspective of time sends a signal. Terra II is our predestined response.

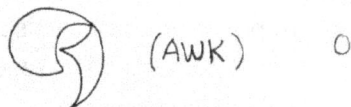 (AWK)   0

 (BEE)   1

 (CHEE)   2

 (LUNG)   3

(WANG)   4

 (DO)   5

 (PIE)   6

1. The comet Starseed is signal to leave the womb-planet Earth.

(This is the English translation of the 1st Transmission)

Terra II: A Way Out

This book presents the Starseed projection – a practical plan for sending a starship city, Terra II, through the center of the galaxy to the farthest portion of the visible universe, to contact and exchange information with Higher Intelligence. Terra II will enclose a miniature replica of earth, within a sphere two kilometers in diameter. The voyage will last over a hundred years in flight time, the equivalent of more than a billion years Earth time. The trip will require the creation and maintenance of life-support systems, improved social structures, and the bio-chemical suspension of the aging process (bio-immortality).

The Starseed project unfolds in three stages:

1. International convocating will be held in the spring of 1974 to work out the technical, economic, and social blueprints for the voyage.

2. In the fall of 1974 the pre-flight crew will start constructing a functional replica of the Star-city in the Antarctic. The technical and cultural procedures for the trip will be developed during this pre-flight period which will last

up to 24 years. The engineered components for the Star-ship will be assembled on Earth island (to be leased from the Japanese government).

3. Before the year 2000, Terra II will be assembled in orbit around the moon and launched towards the stars. While the ship (due to time dilation-relativity), will not return within the lifetime of any living terrestrial, messages will be sent back to inform humanity about the lessons learned from interstellar contacts.

2. Life seeds — "egg-planets" throughout the galaxy. When life leaves the womb-planet it attains immortality in the galactic star school.

(This is the English translation of the 2nd Transmission)

The Goal of the Trip

Life on Planet Earth is of extra-terrestrial origin and of extraterrestrial destination.

Life was seeded on this planet three to four billion years ago by means of nucleotide templates containing the blueprint for gradual evolution through a sequence of bio-chemical stages. Every living creature from the unicellular to the human is a stamped-out unit playing its part in the overall evolutionary design. The role of each living entity is to reproduce, to carry on the electric chain of life, to contribute its body to the humus-top-soil, its genes to the galactic life-mosaic.

Planet earth is one of millions of "egg" structures, a womb on which life passes through embryonic stages leading to the point of birth. The moment of birth comes when life leaves the womb-planet and the protective membrane of the Van Allen Belt. Birth occurs when Life has produced nervous systems capable of deciphering the genetic code, attaining biochemical immortality, and harnessing the energies necessary to leave the solar system.

Evolution unfolds through larval stages necessary to secure itself in embryonic fashion on the planet and mature in preparation for exit from the earth-womb. We were not designed to reside permanently on the Earth.

At this halfway point in the life of our Solar System it is time for humanity to accept its genetic mission. From this moment the only purpose of humanity is to attain bio-chemical immortality and prepare to leave the solar system. We are too big now for the womb. If we continue to push and shove for intra-uterine dominance we shall strangle, still-born.

SYNCHRONICITY AND RECAPITULATION: THE COSMIC SCRIPT

Scientists tell us that the same laws of energy transformation occur at all levels of the cosmos.

The same processes which govern the movement of nuclear particles direct the movement of galaxies. And the movements of living organisms.

We are told that the basic structure of life – the genetic material – is the same in every form of life on our planet.

The applications of this Scientific Philosophy (Psi-Phy) to the behavior of human beings is simple. We assume that our behavior synchronizes with and recapitulates the evolution of life.

As it is with the individual, so it is with the species. Etc.

Human life begins when the sperm (carrying the seed-code of life) travels to, contacts, and fertilizes the egg.

Astronomers and genetic biologists suggest that life came to this planet via the process of "directed panspermia"

(Crick, Nobel Laureate). To understand the direction of planetary evolution it is only common sense to look at the developmental process which occurs after conception.

The organism remains in a prenatal state until it has recapitulated the evolution of the species, attained the maturity necessary to adapt to the current environmental situation, and then is born. To be born is to move in space and time from the enclosed, protected, womb to the "outside" world.

The prenatal embryo is programmed to leave the warm, restricting womb and to deal with the stimuli and challenges of the "external" environment.

The embryo is probably not aware that She has not yet been born. There is, undoubtedly, a vague intuition. But we can assume that to the embryo the womb is the natural environment. Certainly the embryo can have no knowledge of the environmental situation that awaits. The racial, national, cultural, class-caste factors that later come to define "reality" are totally beyond the prenatal expectation.

Starseed suggests that life on the planet exists in an embryonic state. The earth is a womb. The perverse humor of nature delights in polarities, mirror images, in-out paradoxes. Although the womb of the more advanced species is internal, within the body, the womb of the life process on this planet is external, i.e., on the surface of the earth. Schematic electromagnetic drawings of the earth, as viewed from a few hundred miles in space, show the cloudlike Van Allen Belt as a soft, doughnut-like membrane padding within which the earth rests like a delicate egg.

Life is programmed to be born, that is, to eject itself from the planet-womb and begin toddling around the house. The galaxy is the house. Indeed, it is not impossible that the galaxy itself may be a nursery school and that when we have learned what is to be learned in the galaxy (with its millions of life-bearing planets) we may be whirled off to more advanced galaxies to continue our education.

To those who object to such cosmic theorizing, we must quote the renowned astronomer who said: "no matter how fantastic and improbable our wildest visions about the future, rest assured that the truth will be more fantastic and improbable. Even the most far-out Psi-Phy minds of a century ago could not imagine what is now occurring.

Starseed proposes the "earth-as-womb" theory because there are no facts to disprove it; because the available facts confirm it; and because it is in line with the synchronicity and recapitulation laws. The natural laws which govern conception, prenatal growth, birth, and metamorphosis here on the surface of our world suggest to us how events may unfold beyond our little sun-warmed hatchery.

The earth-as-womb theory also recommends itself because it is funny, hopeful, pregnant with future possibility, delightfully impudent to our pompous self-conception. What more entertaining conceit than to suddenly discover that we have not even been born yet.

The current writhings and convulsive spasms of our species which are disturbing the serenity of the planet can be seen as birth pains.

The planet can no longer hold us. We are crowding and fouling the womb. Rational, socialistic plans to reduce population, restrict growth, think small, control free expansion, regulate expression are timid and suicidal. It is in the nature of every form of life to joyously, thoughtlessly expand and multiply. The growth restriction plans now so popular among scientific bureaucrats are anti-evolutionary. It is as though the embryo decides that the only way to *remain* comfortable in the womb is to stop growth. Conserve energy?

No way! Expend to the limit? $E = mc^2$?

The answer of life to its restrictors is always the same, expand! Burst forth! Mutate! Upward! Turn on, tune in, break out! Get born!

Synchronicity. What is happening to you is happening at the microscopic and telescopic levels. It's the same process in the atomic nucleus and the galactic village.

Recapitulation. The first nine months of our prenatal life repeat phylogenetic evolution. Every state is larval, embryonic to the next. The process is not going to stop with the Tanaka government or the Nixon administration. Painful as it may appear to some, we have not even been born yet.

The crew of Terra II will be formed with the loving magnetism of the chromosome. When our sperm-ship breaks through the meteor belt beyond Saturn and bursts beyond the uterine pull of solar gravity, our life will begin. And we shall grow, and learn, and mature. And we shall become star children at play in the galaxy.

THE CELESTIAL VISION

Until now the human being has existed in embryonic state with no answers to the basic questions: how did we get here? Why are we here? Where are we going? The primitive, larval religions have produced theories, rituals, and ethics which concern embryonic survival. The Codes of Confucius, the Ten Commandments. Placental morals.

However despite their technical vagueness, all larval cosmologies share the triple vision. Immortality. Heaven above. Higher Intelligence. The global persistence of a Noah Myth is also premonitory.

Thus we see that Starseed, far from being an heretical novelty, is the current manifestation of the oldest guiding myths of our species, a renewal of our solar heritage. This moment of birth from womb-earth is no time to be pessimistic about human destiny. From the perspective of Starseed we joyously understand that humanity, in spite of momentary distractions of war, dictatorships, plague, has moved directly and efficiently in the direction of its obstetrical mission: to establish physical security on this planet, to develop the bio-chemical and nuclear technology necessary to liberate the nervous system to the higher states of consciousness capable of attaining immortality and communicating with the creators.

The malaise which now infects the world is philosophic. A pre-partim depression. The ancient, transcendental, rebirth visions expressed in the stone-age metaphors of Shinto, Buddhism, Hinduism, Islam, Judaism, Christianity, etc., have lost their meaning. The old religions, let it be noted, kept the entire evolutionary spectrum open. The purpose of life was to contact the Higher Intelligence in heaven and attain immortality. In order to prepare for the voyage, "scientific" organizations of cosmonautical engineers (e.g., the Confucian scholars, the Catholic church) worked out and executed the placental plans thought necessary to prepare humanity for the rebirth trip (e.g., The Eightfold Path, Bushido, The Ten Commandments). A thousand years ago no one questioned the cosmological purpose of life; everyone agreed that the basic goals were immortality and the voyage upward. The Technological Renaissance developed the scientific and technical steps necessary to attain bio-chemical immortality and leave the planet. Unhappily, mankind became fascinated by its scientific accomplishments, trapped into technological addiction, and over-emphasized the larval processes of physical survival, terrestrial power, material skill and comfort within the womb. The heretical and restrictive power of the state caused humanity to forget, momentarily, that the aim of life is to "contact the Higher Intelligence in Heaven," a goal which can now be defined not in the accurate but poetically vague metaphors of prescientific theology, but in terms of a starship launched in the direction of the astronomically defined center of the universe where, according to all statistical probabilities, we shall contact more advanced civilizations and be instructed as to the next stage in our existence.

DOES HIGHER INTELLIGENCE EXIST?

Here are the only questions worth the consideration of any intelligent person. Is physical immortality possible? Does Higher Intelligence exist? Everyone must say a basic "yes" or "no" to this decisive probe. If your answers are negative, either explicitly or by passive, uncaring default, then nothing makes any difference except petty satisfaction of robot comforts for self and family during this brief, pointless existence. If your answers are affirmative then the most exciting, adventurous, and hopeful vista emerges.

CERTAINLY NO OTHER REWARD OR
PROSPECT CAN COMPARE.

Humanity now hungers both for reassurance that ancient fetal aspirations were not in vain and for reminder that there is a purpose to existence on this planet. What sounds like the most far out and impractical fantasy becomes, upon reflection, the only sensible proposal.

Let us examine the logical possibilities.

1. The Starseed hypothesis: Life on earth is embryonic offspring of advanced civilizations existing on other planets within our galaxy which can be contacted by means of electro-magnetic messagery and Time-ship exploration. (the

Pre-Columbian aspirations are now seen as premonitory.)

2. No Higher Intelligence exists beyond the gradual accumulation of scientific knowledge. Life is a unique development on earth. There is no advanced life in other solar systems and no genetically pre-programmed higher levels of awareness within the nervous system.

No matter which of these hypotheses one aesthetically prefers, the fact remains that the best investment for the human race, the most exciting, inoffensive way to pass our time is to assume, pretend, gamble that there is a Higher Intelligence, to develop an immortality pill and to organize an all-out search for the Star-school. From the history of science we learn that the only way any new energy is discovered is to look for it. Indeed, we tend to find whatever we look for. To paraphrase Voltaire, if the Higher Intelligence does not exist it is time to invent It.

When five thousand persons from every country assemble to live and work together in the dedicated search for Immortality and Higher Intelligence the results, even if negative, cannot fail to be amusing and instructive to the race. There is simply nothing better to live for.

Since there are about one hundred thousand million stars in the Milky Way system this means that some thousand million stars must have planets in the appropriate condition to support long-term evolution.

It may be argued that the fivefold reduction which we have already made in the original estimate is not 5 but 5000 times wrong. In this case we conclude that there must be still a hundred million stars in the Milky Way with planets which could support organic evolution.

When we consider the wider aspects of the cosmos as a whole the situation becomes even more dramatic . . . Our estimates lead us to conclude that in the observable universe there are probably some trillion stars possessing planets in a suitable condition for the support of organic evolution.

Sir Bernard Lovell, Director,
Jodrell Bank Radio Telescope,
in *The Exploration of Outer Space.*

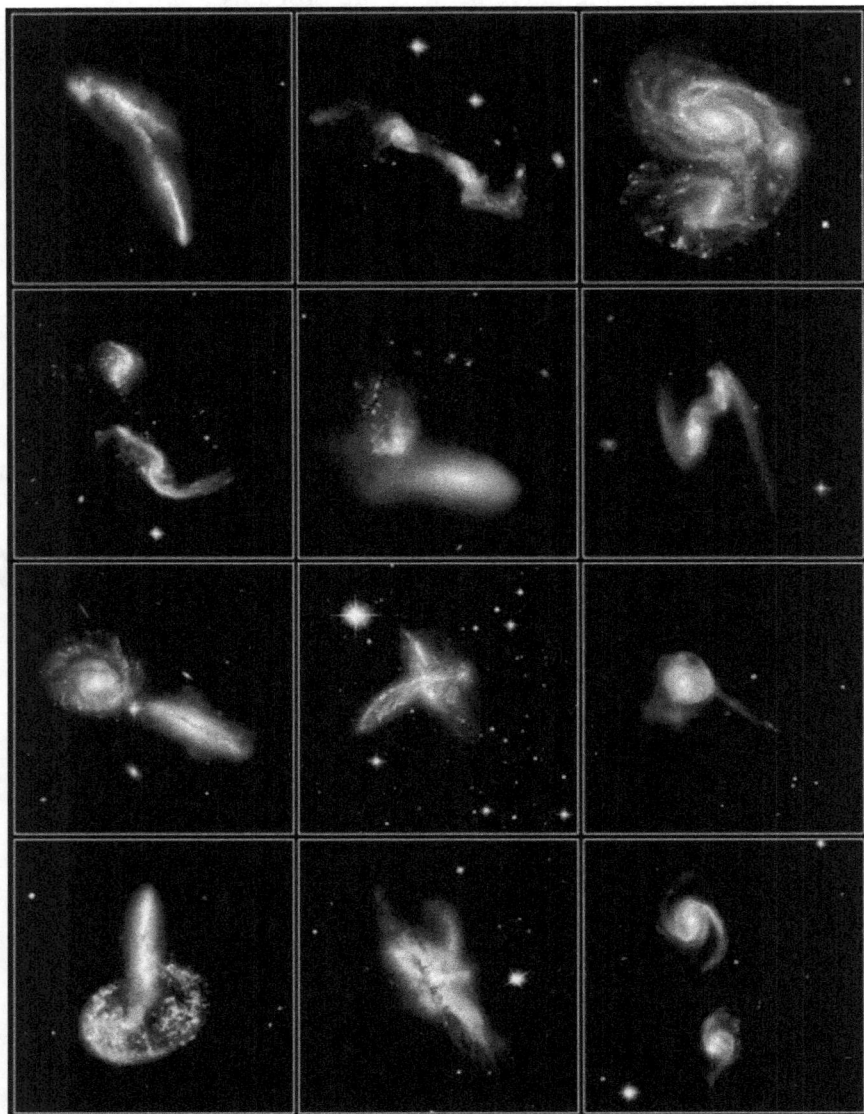

Pictured here are some of the types of galaxies which
Terra II will visit en route to the center of the universe. It
is probable that some of these galaxies are inhabited by
intelligent beings who have formed galactic federations and
interstellar communication networks.

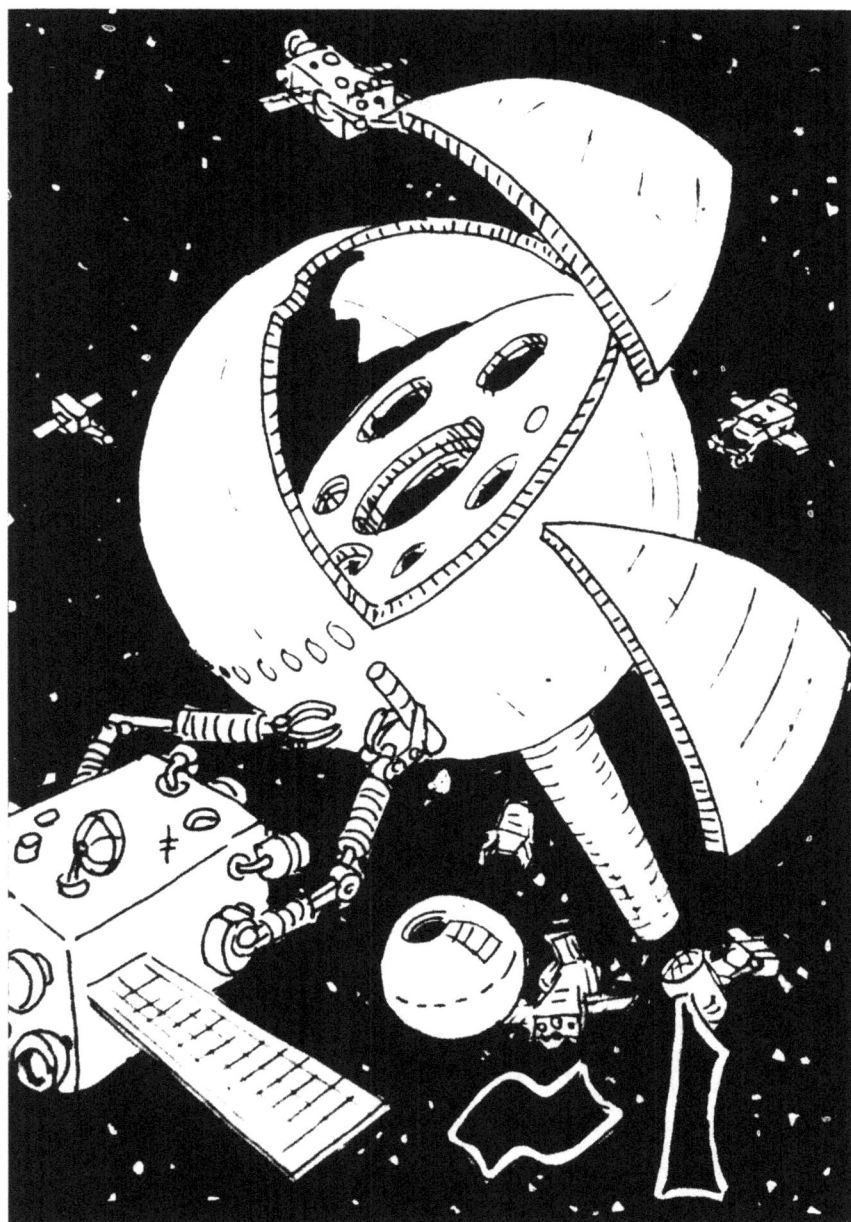

3. When the embryonic nervous system can decipher the genetic code, it receives instructions for leaving the earth womb and contacting higher intelligence.

(This is the English translation of the 3rd Transmission)

Has Life Been Seeded On This Planet?

According to the ancient, pre-scientific religions, extraterrestrial, celestial Creators introduced life upon the earth.

Scientific theories of evolution improved the anthropomorphic concepts of *how* life developed, replacing the crude engineering notions of the seven-day god with genetics. But biological philosophers and geneticists have failed to explain the origin of life.

Darwin concluded that, in the dim past, there must have been some single, primitive form of life from which everything arose, and logically asked, whence came that original species? In a recently discovered letter, thought to have been the last that he dictated and signed before his death, Darwin said the knowledge of that time (1882) was so meagre that any serious attempt to explain life's origin would be premature . . . The Principle of Continuity, he wrote, "renders it probably that the principle of life will here-after be shown to be a part, or a consequence, of some general law."

Science has not yet produced any evidence or theory to disprove legendary and biblical accounts that a celestial Higher Intelligence is responsible for life on the earth.

Indeed, there is a mounting wave of evidence and conjecture that provides the technical explanation of how the Higher Intelligence managed to do it.

GALACTIC IMPREGNATION

Panspermia, a notion of Svan August Arrhenius, states that the templates of life arrived in the form of microscopic spores which escape from donor planets by means of air currents of volcanic eruptions. As Sullivan points out, "The trouble with the panspermia idea was that bacterial spores . . . Would probably be killed soon after they left the protecting envelope of air around the (donor planet) by ultraviolet rays from the sun."

Meteorites provide a more plausible engineering explanation for the arrival of living matter on our planet. It is estimated that more than 1000 meteorites land on our planet every year. "At the microscopic level there is a constant rain of material . . . It has been estimated that this dust falls at the rate of about 1000 tons a day."

On the night of May 14, 1864, a meteorite fell around the village of Orgueilz, France. Twenty portions of the object were recovered which ranged in size from apples to pumpkins. In composition the objects resembled peat, easily cut with a knife, soluble in water. "The Orgueil meteorite has probably been more elaborately studied than any other chunk of material on earth." These analyses have indicated that the meteoritic material contains hydrocarbons and highly varied, complex and baffling organic molecules, including some similar to "Cytosine, which is one of the bases of nucleic acid." On the basis of the evidence that organic materials have reached our planet within meteorites, the hypothesis that life on earth has been seeded from extraterrestrial sources seems, at the present time, more realistic than theories of spontaneous generation.

The philosophic and practical implications of the Starseed Hatchery hypothesis are fascinating. Our home base, our parental source, is extraterrestrial. From the genetic perspective of the seventh neural circuit, we realize that life originated elsewhere in the galaxy. Directed panspermia. Our planet was fertilized by us. We came as sperm, impregnated the planet and have been seeded on planet earth to incubate. When we have evolved to the point where we understand our origin and can locate our "celestial parents," we are ready to emerge from the prenatal incubation and make our way in a mature state. During the first three billion years of our planetary incubation, we have passed through the fetal states necessary to produce a bilateral nervous system capable of: 1. Deciphering the secrets of life which is encoded in DNA; 2. Discovering the cause of the aging process; and 3. Developing the technology necessary to return to the galactic center where we shall find our "celestial parents."

HOW WERE WE PLANTED HERE?

After winning the Nobel Prize for helping to discover the structure of DNA, the master molecule of life, what does a scientist like Francis Crick do for an encore? He tackles something even bigger. With Leslie Orgel, of California's Salk Institute, Crick has now taken on the mystery of the origin of life. Writing in *Icarus*, a monthly devoted to studies of the solar system, the two scientists theorize that life on earth may have sprung from tiny organisms from a distant planet – sent here by spaceship as part of a deliberate act of seeding.

This bizarre-sounding theory, called "directed panspermia"* by its authors, results partly from uneasiness among scientists over current explanations about how life arose spontaneously on earth. Crick and Orgel note, for example, that the element molybdenum plays a key role in many enzymatic reactions that are important to life. Yet molybdenum is a rare element, much less abundant than, say, chromium or nickel – which are relatively unimportant in biochemical reactions. Thus, because the chemical composition of organisms must reflect to some extent the composition of the environment in which they evolved, the authors suggest that earth life could have begun on a planet where molybdenum is more abundant.

~•~

* From the 'panspermia' theory of Swedish chemist Svant Arrhenius, who suggested in 1908 that living cells floated haphazardly through the universe, bringing life to desolate planets.

~•~

Crick and Orgel also ask why there is only one genetic code for terrestrial life. If creatures sprang to life in some great "primeval soup." as many biologists believe, it is surprising that organisms with a number of different codes do not exist. In fact, Crick and Orgel say, the existence of a single code seems to be entirely compatible with the notion that all life descended from a single instance of directed panspermia.

The seeding of terrestrial life could have been carried out by a civilization that was only slightly more advanced than man is now. In fact, Crick and Orgel estimate, man within a few decades will have nuclear rocket engines that would enable him to conduct a little panspermia of his own. Using such rockets, it would be possible to reach planets orbiting around any of thousands of stars with spacecraft carrying microorganisms, such as dormant algae and bacterial spores. Suitably protected and maintained at temperatures close to absolute zero, the organisms could be kept alive for a million years or more.

Why would man, or some distant intelligent beings, ever launch a panspermia project? To demonstrate technological capability, say Crick and Orgel – or, more probably, out of "Some form of missionary zeal."

– *Time*, September 10, 1973

This article from *Time* magazine illustrates and clarifies the meaning of the Starseed Transmission.

1. Every day, scientists working in different specialties around the world present findings which support the general hypotheses of neurologic and Starseed. But these isolated discoveries are usually understood only in narrow context and the implications reflect a thin bureaucratic perspective. Scientists now relate their findings to commercial production, military application and politically popular goals. Almost every scientist is supported either by government or industrial funding. The scope of both the experimentation and interpretation is thereby limited. The committees of Starseed scientists, on the contrary, will be motivated to expand their vistas in preparation for the prospective dialogue with Higher Intelligence. A quantum jump in the level of scientific activity is to be expected. Terra II scientists will not have to justify their work to congressional committees.

2. Crick and Orgel suggest "technological pride and missionary zeal" as motives for "Directed panspermia." This ethnocentric bias suggests the dilemma of the Terra I scientist.

The Starseed answer to the question of motivation: it is in the biochemical genetic nature of human destiny to attempt to communicate across interstellar space. As it is the programmed nature of the flower to send its seed to the wind. The galaxy is hooked up into a communicative network. Starseed assumes that human beings are part of the galactic life-chain and that we operate according to the same principles that spin the electron and send starlight across the lightyears. If pressed to choose psychological motives, instead of "pride" and "zeal" we would cite curiosity, hunger for contact, love, compulsion to expand and grow.

The explanations of Crick-Orgel and of Starseed are equally subjective. Expressions of faith. We operate on the hypothesis that such motives as love, gratitude, interspecies unity are more likely to produce the evolutionary future that we hold precious (and amusing).

4. There is no choice. Life must leave the womb-planet to survive and evolve.

(This is the English translation of the 4th Transmission)

The necessity for the voyage

Evidence has been cited that Higher Intelligence probably exists on planets most likely to be located in the direction of the Galactic Center. Evidence has also been summarized that life on this planet was probably seeded from organic molecules transported from more advanced planets.

The purpose of life, it has been suggested, is to join the interstellar life network.

If the Starseed transmissions are hallucinations it does not matter. Since they are the most logical and practical and optimistic hallucinations available, they can be accepted and acted upon until more amusing, hopeful hypotheses come along.

We organize the voyage to Galactic Center as a concrete act of visionary love. To demonstrate to our galactic parents and ourselves how good we can be. For five thousand people selected from every nationality to live together for eight years to prepare for the voyage will be in itself the greatest achievement of human history, a fascinating and profitable experiment of practical intelligence will have been conducted.

Once Terra II is launched into time those remaining on Terra I will inherit a collective vision and a social model which could produce a new spirit of reconciliation and renewal.

In addition to these comforting logical reasons for the voyage we must face the fact that we have no choice. The mutation from terrestrial to interstellar life must be made, because the womb planet itself is going to blow up within a few billion years.

(It is about time that we abandon our blind commitment to embryonic time spans and begin to think in stellar terms. There is no reason why anyone reading these lines cannot live long enough to make the star-trip, move into Einsteinian time and watch terrestrial centuries spin by like minutes.)

Blunt common-sense trust in the wisdom of the DNA code should convince us that life wouldn't have gotten into this planetary crisis without having figured out a way out. Planet earth is a stepping stone on our time-trip through the galaxy. Life has to get its seed-self off the planet to survive.

The few billion years left before the solar explosion engulfs earth may seem like a long time by embryonic-terrestrial estimates. To those travelling near the speed of light a few billion years becomes the span of a few seasons.

There is another compelling reason for the voyage. It is in the evolutionary cards. Our presumptuous culture-bound minds cannot really continue to deny the facts of evolution. The mutational direction is clearly marked. In the beginning we were aquatic creatures. Then amphibian. Then terrestrial. Next stellar. Can fetal humanity careen along the pathways of social-conditioning ignoring the fact that we are an unfinished larval form in process of mutation? Does not every mythic intuition combine with current scientific evidence to point us to the stars? Do not the UFO sightings, however hallucinatory or hysterical, bespeak some cellular expectation?

We have no choice. We are genetically propelled to move onward and upward.

This photograph shows whole systems of galaxies
scattered across the night-time sky.

Terra II ... A Way Out

5. The fused interconnectedness of the crew will create a new level of energy and intelligence.

(This is the English translation of the 5th Transmission)

Intellectual Fusion

A skeptical reaction to this proposal: since human beings have not been able to cooperate and establish a harmonious use of resources in the past, how can we expect them to reform, cooped up inside a miniature world which, like a pressure cooker, may generate explosive pressures of disharmony, etc.

There are several hopeful answers to this realistic objection.

The population of Terra II is both self-selected and group selected. We shall offer the galactic intelligence our best hopes, our best energies, our best people.

The Terra II population will be highly motivated, united in the greatest enterprise that humanity has ever initiated. Human beings in the past have been caught in webs of ambivalent and conflicting larval motivations. Class, caste, race, nationality, personality, family. The Terra II population will be intimately harnessed together in a conscious evolutionary process.

The most promising asset of the flight will derive from the intellectual freedom which will lift the level of cognitive ability to a new metamorphosized level. At the present time the human species has produced an astounding number of scientific and creative achievements. There is a large population of evolved post-larval persons living around the globe. They are separated from each other. Each person and each group isolated in prenatal social-political structures which, on the one hand, support them, and on the other hand limit them to the restricted visions of the larval political bureaucracy. Scientific goals are deliberately short-sighted to appear practical: cure of cancer, military hardware. Social scientists in particular are limited in proposing solutions to human problems because above all they must not rock the boat. Etc. Institutes which bring scholars and scientists together, such as Rand, Hudson Institute, Salk Foundation, while they encourage cross-disciplinary contact, can do little to motivate global evolution.

The Starseed project can be seen as an enormous process of intellectual nuclear fusion. Five thousand of the most brilliant human beings will assemble on a time-ship. In order to survive they will have to produce new and successful solutions to all of the ancient human problems. The situation will require creativity. At the present time the chemist and the psychologist work at their separate laboratories and then drive home to the suburbs. There is no pressure to integrate discoveries into daily life, to cross-fertilize on a survival basis.

Terra II will more than justify its existence as an incubator of scientific and artistic discoveries in every aspect of life. Each scientific and scholarly discipline will have its research facilities aboard. The results of the ongoing research will be relayed continuously back to Terra I.

It is likely that within years the star-ship will make radio contact with other civilizations. As Terra II leaves the powerful field of solar radiation its signals will be clearer to other scanning receivers. These first contacts with Higher Intelligence will inevitably increase our understanding of physical-chemical-biological processes which in turn will be transmitted back to earth.

Today scientists play at the game of science. In addition to being scientists they are democrats, golfers, adulterers, players in low-level creativity. The novelty and continual pressure for innovative solutions on Terra II will guarantee a higher state of consciousness, an attainment of new levels of neural effectiveness.

Terra II ... A Way Out

6. TERRA II WILL NOT TAKE FROM TERRA I, BUT WILL CREATE A NEW LEVEL OF INTELLIGENCE AND ENERGY.

(This is the English translation of the 6th Transmission)

THE PRICE OF HEAVEN

The Terra II voyage lasting hundreds of flight-years and billions of earth years will cost less than five percent of the current annual global military budget for the next twenty five years.

The Starseed project is self-supporting. International in scope, it cannot ask or accept financial support from any country. Nor can it be under any outside political control.

Starseed is, of course, a non-profit enterprise. No member of Terra II will receive any funds from Starseed for compensation. Starseed will pay outside companies who manufacture equipment and perform subcontracting tasks.

The cost of the Starseed project will be 250 billion dollars. This sum will be raised over a 25 year period by means of: 1. Donations; 2. Research contracts; 3. Sale and lease of media rights.

The entire project will cost less than four dollars a year for each person on the earth.

FINANCING THE VOYAGE

1. *Donations*: Starseed finances will be handled by a trust which will receive donations. The enthusiasm of donors is explained by the general reaction that Starseed is the first totally sensible idea ever suggested for the use of human resources. By 1976 it is expected that a thousand million dollars will be received. Another billion dollars will come in smaller contributions. As the project becomes tangible and concretely realizable the flow of donations will increase geometrically.

2. *Research contracts*: the Terra II crew will include more than one thousand gifted scientists designing new solutions for every aspect of human life. The patents on discoveries made by the various research groups will be licensed or sold for more than one billion dollars a year.

3. *Sale and lease of media rights*: the Terra II crew will include more than a thousand of the most talented musicians, artists, writers, scholars, film-makers, architects, designers, artisans, etc. The activities of the earth-city will become of great interest to the inhabitants of Terra I. The creative productions of the crew and the media rights to news chronicling of Starseed events will bring in more than one billion dollars a year.

The Starseed trust will, in addition, sell long-term rights to the inventions, productions, and creative works of the population of Terra II for the years after the launching. As the years of the voyage pass, the value of Terra II productivity will increase to the extent that the output will become literally priceless. Terra II will have already paid its own way, recompensed Terra I for the necessary supplies and equipment before launching. During the centuries of the voyage Terra II will be freely donating back to the home planet the fruits of its discoveries.

It is probable that over a period of years the non-profit Starseed trust would, under present capitalist procedures, end up owning most of Terra I. After leaving the Solar System all of Starseed's assets will be contributed to the welfare of Terra I.

Even though the inhabitants of Terra II become a new mutant race, their familial indebtedness, gratitude, and concern for the fetal planet will continue. The very existence of Terra II as a higher-conscious evolute from Terra I will, it is expected, serve as a model and as an inspiration, and, perhaps, even as a conscience for Terra I to emulate.

Terra II ... A Way Out

7. THE HEAVY STRUCTURES WHICH SEEM TO OBSTRUCT THE FLIGHT ARE NECESSARY TESTS OF STRENGTH AND READINESS. LIFE IS A UNITY. EVERY LIVING CREATURE YEARNS FOR THE VOYAGE TO OCCUR.

(This is the English translation of the 7th Transmission)

A REALISTIC AND FRIENDLY LOOK AT EVIL

Good and evil refer to bi-polar processes which are necessary in the evolutionary process.

Good, as defined biologically, refers to those events which enhance evolutionary growth, contribute to the harmonization of the All-life process.

Evil, the vice versa of live, refers to those events which inhibit evolutionary growth, enhance one facet of the biological unity at the expense of the whole.

In the narrower context of this project, good is that which leads us to look to the stars, immortalize, mobilize our energies to send a living signal of our love back to Galactic Center.

Evil is that which keeps us down, impels us to engage in competitive domination of other life forms.

Evil is as necessary as good.

Evil is needed to test us. The evolutionary process is not simple. Existence is not guaranteed. Life is a struggle towards the light. Delicate, vulnerable seeds tossed out into space, drifting across light years, we fall on this planet, a globe of molten rock, boiling in the solar incandescence without an oxygen atmosphere. Once rooted, living creatures spread over every inch of the planetary surface struggling to survive. Species arise and are destroyed by the fierce amniotic competition. Stronger oaks shade out weaker trees. The weak are eaten to enhance the strong. In addition to these low-level Darwinian vulgarities there seems to be a factor that operates to choke off the advance of species which become overdeveloped in terrestrial skills.

It is a sobering fact that every advanced civilization in human history has eventually fallen to destruction.

There is a force, a timing vector which has operated to slow down evolution, to keep human vision narrow and earth-bound. For example: the theory of extraterrestrial worlds was promulgated by Thales, Anaximander, Pythagoras, Xenophanes in the sixth century B.C., a period miraculously explosive with the cosmic visionaries. Politics, both local and geocentric, crushed these perspectives for two thousand years until the renaissance initiated the renewal of which we are the product.

We can be sure that now as we begin to organize the startrip for which we were predestined, strong gravitational forces, national and geo-centric, will seek to prevent the voyage.

This is natural and to be expected.

Unrelenting conservative pressure has always opposed each evolutionary advance. The mutant is always feared and hated because the larval form currently dominant instinctively resists change. The mutant must prove its strength and superiority against the fierce challenge of the earlier form.

Starseed does imply a mutation of the human being. Terra II is leaving Terra I. This evolutionary separation will not involve force or violence. The mutation is one of intelligence and love. It is the task of Starseed to demonstrate to the Terran I society that the voyage into time will not in any way injure planetary existence, but on the contrary, provide knowledge, entertainment, and spiritual meaning. Starseed costs nothing and hurts no one.

Thus evil is neutralized by good, space by time, gravity by levity, fear by hope, past by future.

THE PTOLEMAIST CONSPIRACY

It is naive to expect that everyone, for every sort of reason, rational and irrational, would hope that there are wiser folks around. In addition to the yearnings for Divine Parental protection, for celestial Big Brothers, there are the eminently logical desires to learn, from our more advanced cousins, lessons which will teach us how to live and grow.

Will it come as a surprise to learn that there are some among us who do not want there to be intelligent life in the galaxy and who will use the considerable influence they possess to prevent any attempt to contact Superior Intelligence? We recall that Giordano Bruno was burned at the stake for suggesting that interstellar space contained inhabited planets.

There are also some among us who are bored with the amniotic level of mentation on this planet and look up in hopes of finding someone entertaining to talk to. Curiosity about adults is another delightful and usually harmless avocation.

It is most revealing to determine the motives of

those whose role it is to deny the possibility of Superior Intelligence and to oppose its search.

They obviously think that they have something to lose. Both closet and overt Ptolemaists apparently expect that Higher Intelligence would be hostile to them. Maybe they are right. In addition to the invasion-from-Mars-communist-conspiracy mentalities, there are many others who probably imagine that the Superior Species would come on like smug Harvard professors, meddling with morality, and making middle-class people feel inferior.

We face here, again, the classic problem (which we pray that the galactic grown-ups have solved) that the smart people ask questions and listen while the stupid are certain of the answers.

Carl Sagan has provided premonitory warning based on press reaction to the Mariner IV fly-by of Mars in 1965. The mass media know intuitively and precisely what the American middle class wants to hear. The facts of the fly-by were simple. This spacecraft was not designed to search for life on Mars. As the experimenters were careful to point out, the mission neither demonstrated nor precluded the possibility of life on Mars. Why then were the communications media so quick to deduce a lifeless Mars? "I believe," says Carl Sagan, "a partial answer can be found in the response made to the Mariner IV findings by political leaders, by Mr. Billy Graham, and by other American divines – often sure barometers of common attitudes. They were unmistakably *relieved*. Finding life beyond the earth – particularly intelligent life, although this is highly unlikely on Mars – wrenches at our secret hope that man is the pinnacle of creation, a contention which no other species on our planet can now challenge. Even simple forms of extraterrestrial life may have abilities and adaptations denied to us. The discovery of life on some other world will, among many other things, be for us a humbling experience."

Even in conception Terra II becomes a challenging yoga, a relentless intellectual and emotional exercise, testing every one of our social and religious assumptions. A fascinating invitation to specify our highest hopes for human life and beyond.

The nearest spiral galaxy, M31, the great nebula in the
constellation Andromeda. Also shown are two smaller
companion galaxies.

8. TERRA II IS THE SEED-BLOSSOM OF THE ONE TREE OF LIFE.

(This is the English translation of the 8th Transmission)

THE TREE OF LIFE BLOSSOMS

We have stressed the historical harmony of the Starseed vision – fulfilling as it does ancient prophecies and spiritual aspirations.

Physics and astronomy these days inevitably lapse into poetic concepts – death and rebirth of stars, anti-matter, black holes, Shivaic cycles of creation and destruction. Etc. Our most objective scientists begin to rave about infinities and eternities like Sanskrit poets, or high school acid heads. All this confirms the hermetic doctrine that the same laws which govern galactic motion and particle movement also operate at the level of human action. What is above, so below. As with the great, so with the small. As with flora, so with fauna.

One of the most persistent and appealing metaphors for human evolution is the venerable gnostic-kabalistic Tree of Life. While the lengthy astro-engineering pages of this book demonstrate the practicality of the Starseed voyage, it must be reassuring to the less objective that the aspirations and teachings of the gentler religions here find renewal and reaffirmation.

The tree of life metaphor expresses the unity of all living entities. The planet was seeded by nucleotide templates which produce various manifestations of the original unity. All life on this planet can be seen as a functional part of one organism, one world plant. Just as the tree has various components – root, trunk, bark, limbs, leaves, seed-carrying nuts and fruits – so does the tree of life extend around the globe, one unified bio-electrical network.

The aim of the tree of life is to produce a flower; and from the flower a seed. Terra II does not represent the glory of humanity, but rather the inevitable evolutionary product of the life process. The flower has no choice. We have been programmed, stamped out by the genetic code to perform this function as sexual organ, pollen carrier for the tree of life. We have no choice but to carry the signal of earth-life to the stars.

The laws of nature are uniform. The stars move to the same electromagnetic rhythms that cause attraction of man to woman. From botany we learn the basic cosmological lesson. Cross-galactic-fertilization is necessary for rebirth and new growth. Interstellar heterosexuality is the glory of evolution. Interstellar love is the name of the game.

Starseed sends across lightyears its pollinating sperm sphere, miniaturized replica of earth, laden with the treasure of our planet.

This is, of course, the prophetic meaning of the 1960's "Flower child." First, tender innocent blossom of the first post-Hiroshima generation.

The spiral galaxy M104 in the constellation Virgo, this galaxy is oriented so that we see it edge-on, it appears much as our own galaxy would look if it were approached from the galaxy Andromeda,

Terra II … A Way Out

9. Womb-planet earth and Terra II are experiments in the bio-chemistry of love. The successful voyage of Terra II is important to the evolution of the galaxy.

(This is the English translation of the 9th Transmission)

Galactic Feelings of Inferiority

The seminal ideas of Arthur Clarke have prepared us to understand the Starseed Transmission.

We have wondered why Dr. Clarke has not gone on to discuss the human community aspects and, indeed, why he has not started to put the trans galactic plan into operation.

If the trip can be made, why do anything else? What embryonic emotions here on Terra I can hold us back?

Cosmic pessimism? Terrestrial fatigue? Pre-partum depression?

Dr. Clarke has just published a book which throws light on this question. *Rendezvous with Rama* (New York, 1973) poses a Psi-Phy mystery. An enormous space ship enters and parks in our solar system. An American interplanetary craft investigates, wanders around outside and probes the inside of the vessel. The dismal conclusion: the alien ship is tanking up on solar energy. Its crew has advanced so far beyond our evolutionary position that it has not the least interest in the puny activities of mankind.

Will this be the disappointing conclusion to the flight of Terra II? When we reach the outskirts of the galactic confederation will we discover that we are of such low evolutionary stuff that they won't even deign to recognize our presence? Indeed, if a Higher Intelligence has already discovered faster-than-light drive, then we must assume they have already scanned our solar system for interesting developments and found us of no interest. Otherwise, why have they not contacted us? Or have they?

Clarke's book suggests this humiliating parallel. We drive into a service station to gas up. There is an ant hill. While we wait for the tanks to fill perhaps one ant crawls on the wheel. We care not. Even if one ant climbs on to the motor, we care not.

If the ants crawled in file on the windshield and began to spell out $E=MC^2$ would we have patience to understand? Suppose they spelled it out in ant alphabet?

Our most advanced nuclear physicists help us not. They say that our most sophisticated mathematical models are simply mental projections. The physicists of the last generation projected a cold universes, impersonal, hostile, pessimistic. We grant them their right to devise whatever myth they must. But we reserve the right to construct our own science fictions, equally valid.

Physicists are convinced that entropy rules the worlds. The cosmos is winding down. "The universe," says Dr. Dicke of Princeton, "tends in the direction of more disorder, of increasing entropy, so the direction of increasing disorder is the direction of time flow."

The entropy interpretation is a fantasy. A pessimistic

cosmology devised by a generation of scientists who appeared at that prenatal moment in history when the two-thousand-year-old religion was collapsing. When embryonic visions were ending.

As this particular cycle unfolds, energy decreases and structured differentiation increases. It is always the twilight of a culture which creates the greatest elaboration and baroque precision. As solar radiation loses its accelerated thrust, as it slows down and cools, it creates a myriad of finely arabesqued details. Then the reverse movement begins and the universe coils back upon itself, sucked into the intense negativity of the black hole. From which a new cycle even more complex explodes. "A one-centimeter black hole," says Dr. Ostriker of Princeton, "Would weigh 10 grams, or the mass of the earth, I think."

The pessimistic *fin de siècle* romanticism of current physics delights in such gloomy concepts as the second law of thermodynamics and entropic weariness. Such thinking is linear and overlooks the possibility of cycles and rebirths and mutations and metamorphoses.

The neutral facts are that we do not know what awaits us as we approach the center of the galaxy. It is almost certain that we shall contact Higher Intelligence. But the reaction is unpredictable.

Inter-stellar Expectations

The Higher Intelligence may be a Stalin-Hitler computer mentality that will either enslave or eliminate us.

The Higher Intelligence may be necessarily antiseptic and, as a public health measure, spray us away as a dangerous infection.

The Higher Intelligence may be totally indifferent to us. We may slide ant-like right through its radiant belts unknowing perhaps even unobserved, infinitesimal.

The Higher Intelligence may have seeded us, planned for our return, and may receive us with amused, delighted parental tolerance. And what did you learn in school today, little ones?

The Starseed Transmission suggests that the Higher Intelligence may need us to fit into some small but vital part of its immense unified design. We are apparently the long-awaited proof of some experiment in the survival of love on dark barren rocky outpost planets.

The Higher Intelligence may be the fulfillment and radiant embodiment of our most secret yearnings.

We believe that the galaxy is a network of harmonious unity. That even as we cherish the smallest unit of life on this planet so shall we be cherished. We believe that love will energize our voyage, love will guide and love will be the outcome.

But beyond these cosmological aspirations, beyond these religious faiths, we are motivated by curiosity. The answer to these questions awaits us in the stars and nothing can prevent us from searching. The fetal rewards which this planet offers are as tinsel to us. We must seek to find.

"No field of inquiry is more fascinating than a search for humanity, or something like humanity, in the mystery-filled happy lands beyond the barriers of interstellar space . . . As far as we can tell, the same physical laws prevail everywhere. The same rules apply at the center of the milky way, in the remote galaxies, and among the stars of the solar neighborhood. In view of a common cosmic physics and chemistry, should we not also expect to find animals and plants everywhere? It seems completely reasonable; and soon we shall say that it seems inevitable."

– Harlow Shapley, Director, Harvard College Observatory.

Terra II ... A Way Out

1O. THINK OF THE STAR-SHIP AS BODY; THE CREW AS NERVOUS SYSTEM; EACH CREW MEMBER AS NEURON.

(This is the English translation of the 10th Transmission)

THE BODY POLITICS OF TERRA II

One of the interesting by-products of the Starseed project is this: an entirely new society will be formed.

The new community will have to fashion new methods of:

1. Physical survival

2. Political organization

3. Technical manipulation, communication

4. Familial sexual interaction

5. Psycho-somatic expression

6. Neuro-electric, computerized consciousness and communication

7. Genetic evolution, immortality

Decisions about the political dynamics of Terra II must await the deliberations of planning committees, the on-location experiences and experimentations of the pre-flight colony. With these qualifications in mind, we can present the first of a continuing expression of beliefs and aspirations about the political issues, discuss some of the political challenges and suggest some of the possible directions of a harmonious community.

The first priority of the Terra II community is physical survival. The health and physical survival of the crew, human and non-human, must be preserved.

The second priority is complete individual freedom in so far as it does not threaten physical safety or the freedom of any other individual.

The third priority is open communication. Everyone, at all times, has complete access to all information which affects him or her personally or collectively. There will be no secrecy. There will be complete privacy in matters which do not directly affect others.

We deal here with potentially explosive problems which have never been solved by any previous social system. It will be necessary to develop psycho-politics a quantum leap beyond any system previously known to humanity.

The model which the Starseed transmission suggests for the political structure is that of the human body and the nervous system. The population of Terra II is seen as a living unity. Each individual a neuron in continual electronic communication with the collective "brain" and capable of establishing instantaneous communication with any other neuron.

The aim of the political system is not power, dominion, or authority, but accurate communication and coordination.

ELECTRONIC DEMOCRACY. EROTICIZED COMPUTERISM.

The brain of Terra II will be a master computer.

The programming of the computer, designed by committees, will be approved by and responsive to every individual. Decisions which affect the survival of the ship will be made on the basis of factual data and rational consensus can be expected. Decisions about matters of style, comfort, aesthetics will be made in such a way as to satisfy as completely as possible any realistic whim, wish, or preference without infringing on the rights of others.

The society will be so defined that there can be no crime. There will be no money. Greed for material possessions will be limited to aesthetic objects since necessities will be provided for. The only form of anti-social behavior that might pose problems would be destructive violence or coercive force, manifested either by an individual or a group.

Using the model of the social body, individual destructive behavior will be considered illness. Every attempt will be made to avoid using force or confinement to deal with destructive behavior. To the extent that violence is motivated by any frustrated demand that can be realistically satisfied, it is the responsibility of the society to do so. In the case of irrational and uncontrollable violence requiring isolation from the community, every attempt will be made to see that such confinement will not be retributive or punitive, and that the person is given every opportunity to understand and voluntarily select the solution to the realistic problem. Such incidents must be seen as problems of society and not of the individual.

It is recognized that the very nature of Terra II life might produce psychological pressures which do not exist as nakedly on Terra I. Claustrophobic reactions. Anxieties and fears related to the irreversibility of the voyage. Tensions due to high population density and the inability to escape.

Some of these pathologies will have been avoided by

pre-flight selection and preparatory living in the earth replica. Even so, psychological disturbances are to be expected. But also to be expected are advances in psychological understanding and treatment. The social structure which allows for both absolute privacy and close group support will protect the fragile. Those who do break down will be considered as integral parts of the whole, cherished by the knowledge that in the evolution of individual growth, temporary "breakdowns" are sources of future re-integration at a higher level, and that from the genetic perspective every individual represents an important signal from DNA, and plays a role in the evolution of the species. Indeed, individual breakdowns may serve as premonitory messages forecasting trends which may become epidemic or testing the strength and durability of our love.

It is expected that if the entire community shares this feeling of cooperative unity, the percentage of psychotic casualty can be reduced to insignificance. Evidence from China seems to indicate that the psychosis rate can be reduced dramatically by group support. While Terra II swings to the opposite pole by encouraging individualism, even to the point of eccentricity, the general theory that a unified society, tolerant and acceptant, can easily absorb sporadic deviance.

The Terra II society will stress psychological and neurological awareness. The model should be: every person a neurologist. Every Terran II will learn how to understand and direct his or her own nervous system.

The vulnerability of individuals and groups to sweeping epidemics of unconscious motivation will be a focus of constant attention. We anticipate periodic waves of euphoria and depression, irritation and congeniality, energy and passivity, routine and creativity, boredom and excitement.

Constant feedback and attention to individual and group neurology. It is possible that just as daily weather reports now acquaint the citizenry with meteorological developments, so will the publication of daily hedonic indices keep each Terran II aware of the neurological climate.

The human brain is continually receiving reports from every one of the twenty billion neurons and returning relevant information to each neuron. In similar fashion each Terran II will be aware of the myriad processes of the social body, the ebb and flow of temporal rhythm, continual experimentation will be encouraged, one hopes with a sense of humor. During periods of gloom Terrans II will have a chance to participate in hedonic counter-measures, ritualistic, pharmacological, aesthetic. The computers may well be programmed for surprise, mystery, artificially induced change.

It is probable that the human nervous system under these conditions of unity, collaboration, security, stimulation will manifest unimagined assets of creativity and intriguing novelty.

Every aspect of Terra II life will attempt to prevent a Big Brother mentality. Every crew member will explicitly agree, as condition of selection, to function as a free, responsible person, aware of the paradox that freedom requires responsibility.

From our knowledge of human history we can anticipate that the greatest danger to harmony, unity, and freedom will be the development of minority groups which attempt to seize power and impose their will upon others.

Divisive Tendencies

The population of Terra II will be composed of all nationalities and social backgrounds. It would be most naive to expect that upon stepping within the time-ship all of the destructive and divisive tendencies of our species will be automatically left behind.

There will be a natural magnetic tendency for groups to continue their Terra I patterns of clustering. People will gravitate towards their own linguistic group. Cultural singularities will attract those of like habit and custom. Cultural diversity must be preserved and strengthened as long as it does not threaten the safety or freedom of others. There is certainly no wish to produce a homogenous, clean-cut astronaut conformity. Terra II offers itself to the galactic mind as a rich variegated seed-ship reflecting the wondrous range of human difference.

A new Terra II culture will develop. A new language, new customs, new mores. But the earth idiosyncrasies must be preserved as well. Each race, nationality, culture must contribute its part to the mosaic. Every attempt will be made to arrange diet and materiality to preserve cultural origins. Yet basically every Terran II must consider him-herself a mutant, a member of a new time society. Nothing is lost or renounced. All past contributes to the future. All decisions made with the awareness that life is a risk, evolution a gamble, Terra II a magnificent adventure. Fear is the only enemy. Conformity the ever-present seed of eventual robotization.

It is expected that the tremendous range of discovery and novelty, the total turn-on of awareness (each detail of life will be conscious and flavored with discovery) will prevent the re-instituton of Terra I chauvinisms.

But human nature . . . the political planning must take into account every negative possibility. Alert, intelligent anticipation of every eventuality is the key to survival.

Racial and cultural tensions will arise. Some will be tempted to assert dominance and power. Pecking orders and hierarchies are found in every species of fowl and mammal. Sexual rivalries and sperm prerogatives are so built in to the genetic code that they can be expected to appear in Terra II.

Eventually someone will be led to stir up his or her people. To stimulate rivalry.

It may be expected that certain Terra I countries and cultural groups will deliberately infiltrate Terra II with agents whose loyalties will be to the old group. Nostalgia and homesickness will be expressed in xenophobic impulses.

So how can Terra II protect itself from forcible take-over by one group seeking to enslave or exploit others? The history of the white race, in particular, provides such a dismal record of international collaboration that suspicions cannot be totally eliminated.

There can be no professional army or police force. There can be no spy system. There can be no central control room locked and guarded. How then can Terra II protect itself from violent insurrection?

By operating at a higher level of consciousness; through open-ness of communication; awareness of immortality; identification with the genetic goal; and by continued communication with Higher Intelligence.

Terra II ... A Way Out

11. Humanity now possesses the scientific knowledge to implement the ancient embryonic visions of heaven and immortality.

(This is the English translation of the 11th Transmission)

Science and Religion

During the last five hundred years science and theology have emerged as an anti-religious force.

Science and technology have been taken over by the state.

The state has usually been opposed to the spiritual: limiting human aspiration, narrowing life focus to local parochial concerns.

With science as its instrument, the state has co-opted religion.

Religions have become totally secularized, preaching moral codes which may have some social restrictive value but which have no relationship to the original aims of religion, which are to provide immortality and to get mankind in touch with the Creative Intelligence.

It is a long-accepted error that religions should prescribe ethical rules for social conduct. This is the province of the state. The behavioral codes prescribed by a religion should delineate the instrumental activities necessary to get to heaven, to attain immortality, to be free, and to contact the Higher Intelligence.

These are the instructions from the Starseed transmission:

1. To send the flowers of life to the center of the galaxy.

2. To bring together geneticists and biologists who will perform the research necessary to block the molecular signals which initiate the aging process. To attain immortality.

3. To contact the advanced civilizations which originally seeded life on this planet.

Starseed is, and its aims, the up-to-date-old-time religion. The rituals and rules of the "religion" are the technological and social arrangements necessary to launch and maintain Terra II.

More specifically, the rituals and rules are the technological and social steps necessary for: the preservation and evolution of all forms of life; freedom; intelligence; social harmony; pleasure; neuro-technological communication; immortality and conscious evolution.

The conflict between technology and religion was embryonically necessary. Science forces the old religions to a renewal, a reformation not of basic aspirations but of rituals.

The counter-culture of the 1960's was a renaissance of spiritual feelings. A rejection of the materialism of technology. But the "movement" was negative. It rejected technology in a naive fetal return to outmoded mysticism and nature worship.

The hippie movement failed because it retreated from technology. It regressed back to the old maternal religion. In rebellion against Christian patriarchy, it substituted the Pan-seed cult. The future, however, is neither male nor female. The ancient polarities are unified in a new telepathic synthesis. Etc.

The new philosophy is Psi-Phy. The eroticization and sanctification of technology. Technology used in the service of the basic celestial aspirations which are built into our cellular codes. To get high. To trip out. To locate the creator. To transcend the material limitations.

12. THE CREW OF TERRA II SELECTS ITSELF THE WAY THE GENETIC CODE BUILDS UP ITS CHAIN OF ELEMENTS.
(This is the English translation of the 12th Transmission)

SELECTION OF THE CREW

The selection of the crew will be guided by a simple principle. We seek to assemble 5000 of the most intelligent, brave, creative, entertaining, competent, beautiful human beings.

The process of *nucleation* schedules the birth of a seminal idea. There is a bed of coals ready to be fired. We light a few coals. Nothing happens. A few more catch, there is a statistically predictable point of flammability, an exact number of coals which must be lit before the entire works ignite.

Starseed awaits the strong. Terra II cannot offer itself as a therapeutic solution to larval problems of ego, neurosis, dependency, rebellion, loneliness. Etc.

Starseed is assembling an enormous amino acid molecule. Each person is to become part of a human DNA chain. This conscious imitation of the double helix life style could be considered the highest form of worship. We assume a certain kinship with the men and women who devoted their lifetimes to constructing the Egyptian pyramids. The Gothic cathedrals. The Amaratsu temples at Ise, the Taj Mahal, the Mayan temples. The DNA code is heliotropic. It looks to the sun for energy and guidance. Perhaps we have been energized by solar radiation to make the galactic connection. In any case, there is stuff enough here for any cosmology addict.

Starseed attracts the free. Freedom means the acceptance of responsibility. Thus we seek the most responsible. Those with the highest consciousness. Such people are dependable. They can receive, they can give.

To say that Starseed seeks the responsible is not to invite the grim serious. The responsible person can be merry, flexible, honest. Some will be volatile, impulsive, aggressive. To build a body, vegetative or animal, the DNA code uses a wide variety of elements. There is no human characteristic which will not be needed to construct the great molecular structure necessary to bring life to the stars.

Every human activity, profession, vocation, avocation, skill, talent, genius will be required. Starseed will attract and select the best. Selection of the crew is a global mating dance. Invitation to the fittest. While age *per se* is not a criterion of stellar selection, it is expected that the crew will comprise for the most part those younger than forty – the demands for flexibility and radical change in life style, etc. The best young astronomers, physicists, musicians, agriculturists, managers.

We are repeating the experience of the European colonists three hundred years ago who selected the crews to cross the Atlantic and settle the new world. And learning from their mistakes.

Selection will avoid overloading on obedient, dependent, bureaucratic types. While it is true that Starseed will support all its members, total initiative responsibility must be assumed by each Terran II. Many of those selected will be rich (since wealth is positively correlated with intelligence and vigor) and they will bring their assets to the venture. Most of those selected will be the top persons in their field and will contribute their talents. During the first pre-flight years half the crew may be sending back from replica city to Terra I the results of their productivity. The writers their books and articles; the musicians their new compositions. Inventors, scientists, engineers, film-makers, actors. The average income of the crew members, before selection, will be over $50,000 a year. If half the crew continues to produce at this level after joining the crew, the annual income thus accruing will total more than one billion two hundred and fifty million dollars.

We mention these practical trivia to suggest the enormous potential: 5000 of the world's most productive talents harnessed together in common cause.

Preliminary discussions inevitably bring up the question of chauvinistic selectivity. Are there any races, cultures, castes that we do not want on the trip? There are strong arguments for limiting the crew to favored groups who have demonstrated peaceful cooperation. If we include North Koreans can we include South Koreans? If we include Irish protestants dare we bring along Irish Catholics? Would it not be more honest to admit that certain human groups have consistently not been able to get along with each other and frankly select those groups which do have a history of congeniality?

No one can impose subjective preference on others; such decisions must await the actual face-to-face meetings of the selection committees.

However, the basic belief of the Starseed project is this: all life is sacred, all life a unity. Every attempt must be made to select a wide and varied range of life forms. Survival necessities must be accepted. The general rule must be that we select the strongest, most viable, superior stock from each species of life and from each human group. We expect that every human culture will be represented by their wisest, strongest, and best. Selection will be primarily based on occupational, professional criteria. We are selecting mutants whose basic loyalty is to the future of the new species rather than nostalgic commitment to past embryonic social forms. Representation will be proportional not to the number of persons in any racial or national group, but rather to the needs of the new society.

Our model is the genetic code and its technique of selecting the elements necessary to build the chain of life. In this context it is of interest to note that the basic unit of DNA is four. Adenine, Guanine, Cytosine, Thymine. It has been suggested that the volunteering and selected unit of Starseed could be four. That is to say that instead of persons volunteering individually, people would be encouraged to volunteer in units of four persons who have been able to commit themselves to spending the rest of their lives together. While the majority of the crew might volunteer separately, the presence of four-person units might provide a stability and solidity at the onset.

THE HUMAN ALCHEMY OF THE VOYAGE

At the time of this writing the Starseed molecule consists of two men confined in a maximum security prison and two beautiful, energetic women totally dedicated to their release.

To the superficial observer it might seem that four people so situated are in no position to propose what is obviously the most grandiose project in the history of our species.

To the person with a sense, either intuitive or scientific, of the dynamics of energy electro-magnetic or biogenetic – this situation will be seen as precisely the most basic, and pregnant. The energy equation is perfectly balanced in-out, male-female, enslaved-free. Two men within and two women without. We have here, of course, an elemental pattern analogous to the structure of the atom; a human duplicate of the fourfold amino acid situation from which all life on this planet has evolved.

This small, but perfectly balanced structure is capable of generating, handling, and expressing enormous energies. This quartet structure is the human replica of the four purines and pyrimidines that comprise DNA. It seems logical that the human social seed molecule would replicate the basic model of life.

The energy generated in this molecule expands outward in geometric ration. The replications and combinations can create infinite variety.

We are forming a time structure, not a space structure. The basic force and form is consciousness, not material. The material structure crystallizes around the energy which can be called love.

When the first DNA molecules began the process of expansion, all the elements necessary for growth were on the planet. The same situation exists today. All the human elements are here, scattered around the planet. All that is required is the hooking-up process – which will be initiated by shocks, resonating signals. It is our responsibility to provide this stimulus. The process of energy linkage is already taking place. This essay is one form. There will be others.

THE HINGE OF EVOLUTION

Q: Then there are other worlds such as this?

A: Many, as this planet used to be. Many more, equally beautiful, but different.

Q: Inhabited by people such as us?

A: The life network on most planets is similar to yours. On some planets the evolutionary process gets distorted, side-tracked, and blocked. Many have reached and passed the point of evolutionary birth.

Q: (Puzzled look.)

A: The life process on every planet reaches a point in its development when it ceases to be a blind product of instinctual fetal wiring. And takes responsibility of its destiny. At that moment it decides for itself which of the various paths it will take – and in every case the creators reveal enough of the plan for even the least intelligent of their children to make the right choice. Some do. Some do not.

Q: Why is it necessary for this decision to be made? Why have the creators made it possible for so many to lose their way?

A: Freedom. Do you understand? Freedom of will. Freedom of choice. Freedom is the principle of the life process. Freedom means responsibility for the decision. Motion is relative. Freedom, progress towards the creator cannot exist unless the other alternative exists.

Q: Our planet seems to have turned away from the stars.

A: The evolutionary process always hinges on the fifty-fifty probability. Life on this planet is obviously approaching the moment of natal decision. It is always a small minority which carries the process forward. Sometimes the balance has been held by only four, who by some miracle – by some special quality implanted at the very beginning – achieve a spiritual development in advance of their contemporaries by as much as a hundred generations. This is especially true of your planet. Life processes on other worlds which began at the same time have progressed far beyond your present state. A few others have dropped out of sight completely. Do you understand?

Q: (Preston nodded.) I am beginning to.

– Paraphrased from Noah II by Roger Dixon, Ace Books, N.Y., 1970

THE GERMINAL IDEA

Starseed is a germinal idea.

The Starseed transmission contained in this book is a signal aimed at our nervous systems.

The hope is to shake up our synapses. Shock us into mutation. Start our axons and dendrites flashing new wig-wag, semaphore beams. Turn our radar antennae towards the center of the galaxy. Not towards the sun because we are the sun.

This message is a rocket cutting right now through the atmospheric magnetic belts that protect us from galactic radiation. Directed panspermia. This book will explode seeds on our surface, encrusted with habit. The particular form that evolves from the intersection of this-book-and-you is unpredictable and unique. Your neural circuits may be too robotized to react. It may be your fetal role to transmit the book to someone else.

The Starseed transmission is obsessive and irresistible, a message of love, offering the logical and scientific path to immortality. "Heaven," Higher Intelligence.

The precision of its evolutionary perspective provides comfort and hope.

The message is that humanity is just now emerging from its larval immaturity ready for the real adventure of life. As the butterfly looks down with loving nostalgia upon the straining conflicts of the caterpillar, so does the Terra II person observe the politics of earth.

Terra I is doomed. There is no political solution to the irresistible growth within the womb of military dictatorships. As population increases the social pathologies will increase and so will the law-and-order police state control. There is no land left on this planet where libertarian mutant experiments can exist.

If your nervous system accepts the Starseed transmission then you are automatically on the trip. You are Terran II.

The power of fetal religious imprints is awesome. There is no way to reason with or scientifically convince an orthodox Arab, Jew, Marxist, Catholic, etc. For the believer, comforting hope is irreversibly imprinted to the larval imprint. Terra II offers the same comfort in a rational context. This is both a strength and a potential danger to the attainment of the goal.

The Terra II concept can transform human nature even before the time-ship leaves the ground.

In ten years we shall have 5000 of the most intelligent, creative people on the globe living and working together. The reaction of those left behind the membrane will be admiring and envious. National, religious, or cultural politicians will be tempted to exploit this unrest. To send rockets to the Antarctic to bomb Terra II is too irrational. A possible solution will be to form competitive, chauvinistic galactic expeditions. Any country, state, county, religious group can set up its own futique experiment. Just as the whole world, in imitation of Japan and America, now lusts for supermarkets and color television, so will each group long for interstellar projection of their own cultural aspirations. Celestial competition. Spiritual consumerism.

Indeed, the voyage may fail through success. Pre-flight replica cities may create such excitement and progress here on the planet that the goal of leaving the planet may be forgotten. Immortality will charge life with more meaning. Relieved of the fear of death, humanity may rediscover the beauties and excitements of this planet and the infinite unfolding riches of the liberated nervous system.

Once our consciousness is directed to the stars we may immediately begin receiving electromagnetic transmissions which could alter the technical aspects of the voyage. The actual voyage may be postponed until the next evolutionary cycle. No matter. The trip to the center of the galaxy is the oldest, most pervasive, enduring vision of mankind. The voyage will be made.

Terra II ... A Way Out

13. THE VOYAGE WILL LAST FOREVER. THE CREW OF TERRA II WILL BECOME ONE ENTITY.

(This is the English translation of the 13th Transmission)

FAMILIAL RELATIONS

The family has been nuclear structure in the evolution of the human race in its larval states.

Family. The four-generation aggregate of grandchildren, uncles, aunts,

cousins, in-laws. Think of the leaves, branches, sprouts and limbs of a tree. The tree of life.

Family. Think of a four-generational time-ship of people, who live together, mutually dependent for survival, hooked into telepathic harmonious communication.

Family. Think of a complex chromosome in which all the different elements characteristic of the nervous system are bonded together. There is the family financier, the family saint, priest, sinner, psychotic, scholar, warrior, drunk, artist, dolt, healer, etc. Each member of the family plays his/her own role and lives vicariously through the expressions and excesses of each other.

A young Kyoto hippie, who was also a medical intern, once said: "My family is a huge boat sailing through the rough seas of life. At the moment I am the far-out port-side element. The entire family is watching me with concern and also pride that the family is in touch with, part of, the new youth culture. They gossip about me, cluck warnings, some scold and denounce, but they all know that I represent the family in the 'antic' probes of the youth movement. That flank is covered."

The lemming growth of urban culture has fissioned the family molecule. The energy released has powered the subway-highrise grandeur of "city," but has created a swarm of alienated individuals and shaky marriage units cut off from the biological and neurological tree of life. Severed from both past and future.

The paradox of time: we can go into the future only as far as we are in touch with our past.

Terra II is at once the most daring, far-reaching tilt into the future and an ambitious attempt to re-institute, renew the basic forms of the past.

Restoration of family-kinship-village life is one of the cultural goals of Starseed. Terra II is a village of 5000 persons floating at the speed of light through time. There will be continual physical contiguity. Soon, everyone will get to know everyone else. There will also be continuous electronic communication. All group activities will be video broadcast. Everyone will be able to tune in on what others are doing (except for private behavior, which is totally under control of the individuals involved). All such interpersonal broadcast is voluntary. The aim will be to establish the vibratory telepathic network of the beehive, combined with total privacy where desired.

Just as each neuron is in instantaneous bio-electric communication with every other nerve cluster, so will Terra II pulse with signal. Each person is a consciousness probe into the future, firing back his/her experiences for the benefit of the whole.

Sub-clusters will develop and evolve.

Each person has his/her own living quarters. These can be shared and combined. Some couples will mate and remain in monogamous contiguity. Groups who share, at the time, circuit attractions, can move into neighborhood contact. The ten villages (500 persons each) will each attract persons who share a special consciousness interest. Envy, jealousy, possessiveness will be minimized because there is no time pressure. Longevity drains off destructive emotion. Mating and friendship are not one-shot gambles. There is time to live with everyone. Longevity plus neural reimprinting makes possible serial reincarnation. You can be anyone. There is time to learn and experience all the social repertoires. Individuals, couples, groups can symphonize their development over time. And everyone will be learning and experiencing vicariously the life probes of others.

We are discussing the fourth neural circuit which mediates the sperm-egg arrangements necessary to provide a stable environment for the thirteen-year period of growth from infancy to maturity.

Morality is the code of behavior prescription that provides fetal stability and continuity for child growth. Social role, responsibility, virtue. We tread here on tricky turf. The imprinting of the sexual circuit of the nervous system.

Terra II starts afresh.

The psychological cards are shuffled and dealt anew.

We seek a society without rules, sanctions, penalties, shames, social fears.

Everyone will be free. Behaviorally and neurologically. This freedom limited by the freedom of others.

Children will not be owned by their parents. Parents will have no authority over their children. Living in village-like environments where everyone looks after and cares for others, children will grow up in a loving network. Education will be individualized. The earliest lessons taught will be the understanding, care, and operation of the nervous system. The science of neuro-genetics will be emphasized.

There will be no social marriage contracts. If couples wish to live together in monogamy they will do so. Their conjugal contracts will be their own business. Every member of the crew will carry responsibilities for keeping the social system going. There will be no hierarchies of status and money. Thus there will be no economic pressures and insecurities forcing couples to live together. While the magnetic differences between the sexes will be recognized, indeed, worshipped, there will be sexual equality.

For reasons of neurological comfort and enrichment it is probable that long-term conjugal relationships will be effected. The aim is to remove the exploitative and repressive aspect of familial contacts while preserving the positive aspects.

Neurologic, consciousness raising, longevity are three factors which will contribute to social and familial harmony.

Neurologic will make possible reimprinting of the nervous system and sequences of serial reincarnation.

Consciousness expansion, both neurological and external, will keep Terrans II alive, curious, rejuvenating. After a few years of voyaging, Terra II will be able to contact superior civilizations at the rate of one or two a month. There will be no reason to be bored or to hoard experience.

Longevity will drastically change emotional life. Couples can live together with an all-out commitment for twenty years; then reimprint and initiate new relationships of equal intensity for twenty years, then return to the first mate, etc.

There is time to experience everything. Life and experience will unfold in a geometric, exponential rate of extension and intensification. No one has anything to lose. Terrans II will become accustomed to living in a state of Olympian divinity. Time no longer relentless terror but endless cycle of grace.

Over the years close telepathic bonds will link all members of the party. Everyone will be "married" in the sense of shared intimacies with everyone else. The uniqueness of each will be a source of enrichment to others. Cultural and personal differences, instead of dividing, will expand the consciousness of the whole.

Terra II ... A Way Out

14. THE KEY TO IMMORTALITY IS CONTAINED IN THE CHEMISTRY OF THE GENETIC CODE.

(This is the English translation of the 14th Transmission)

LONGEVITY: A PRACTICAL NECESSITY

On a trip of the duration projected for Terra II, the longevity of the ship's inhabitants will be of prime importance. To this end, a major area of research for project Starseed will be focused on the phenomenon of the aging process in the human organism. To stop, or at least retard the aging process is a necessary biological technique in star travel. Certainly the starship city will require a long-lived crew. But equally important will be the functional quality of life among the inhabitants of Terra II during advanced age.

A great deal is now known of the symptoms signaling the body's deterioration in advanced age. Cardiovascular or respiratory problems usually dominate the clinical failings; these are often accompanied by debility, wasting and senility. In the past, pneumonia usually terminated the general debilitating effects of cellular deterioration, but today modern medicine, with its antibiotics, intravenous infusions, cardiac pacemakers, respirators, diuretics, and the like, can often resolve the immediate problem of pneumonia or congestive heart failure, until the body's general level of deterioration results in "death caused by old age."

Labeling the symptoms causing death during advanced age has not led to a thorough understanding of the aging process. Our intellectual evolution is tied intrinsically to the length of our biological life span. Retarding or arresting the aging process will allow our minds to evolve beyond the present limits imposed by the debilitating effects of senility and other age-related restrictions of the evolution of the mental process. Death has been considered a natural consequence of life. As long as this attitude prevails, research into this most vital aspect of our biological and intellectual evolution will remain limited.

WHAT IS KNOWN ABOUT THE AGING PROCESS?

The ninth international congress of gerontology, meeting in July, 1972, discussed the increase in mean age, only to discover that little was actually known about the most inevitable processes of cellular evolution. Frederic Verqar, the Swiss dean of gerontologists, proclaimed that, "old age is not an illness, it is a continuation of life with decreasing capacities for adaptation." The main crippler and killer, arteriosclerosis, is not a necessary accompaniment of aging but a disease state that increases in incidence with age; the same is true of cancer. If we could prevent arteriosclerosis, hypertension, and cancer, the life span could be pushed back closer to the biological limit, if such a limit in fact exists.

That genetic factors are important in longevity has been reinforced by several recent studies. A census of the Ecuadorian village of Vilcabomba, taken in 1971 by Ecuador's National Institute of Statistics, recorded a total population of 819 in that remote Andean village. None of the 819 were over 100 years old. The proportion of the population in this small village over 60 is 16.4%, as contrasted with a figure of 6.4% for rural Ecuador in general.

A team of physicians and scientists from Quito, under the direction of Miguel Salvador, has been studying this unique population. Guillerno Vela of the University of Quito, the nutritionist in the group, finds that the average daily caloric intake of an elderly Vilcabomba adult is 1,200 calories. Protein provided 35 to 38 grams and fat only 12 to 19 grams; 200 to 250 grams of carbohydrates completed the diet. The contribution of animal protein and animal fat in the diet was very low. By world nutritional levels, the elderly Vilcabombans exist on a substandard, even subsistence diet. By extrapolation, the rate of centenarians stands at 1,100 per 100,000 in this tiny village, as compared with 3 per 100,000 in the United States. Of particular note was the fact that the old people of Vilcabomba all appeared to be of European rather than Indian descent.

Another geographical area inhabited by a high percentage of centenarians is the land ruled by the Mirs. The country of Junqa is hidden among the towering peaks of the Karakorum Range on Pakistan's border with China and Afghanistan, and is one of the most inaccessible places on earth, according to S. Magood Ali, a Pakistani nutritionist who has surveyed the diets of 55 adult lamas in Hunza, the daily average is 1,923 calories: 50 grams are protein, 36 grams fat, and 354 grams carbohydrates. Meat and dairy products account for only 1% of the total. Again, this group of long-lived people subsist on a diet of barely life-supportive quantity and quality, because of their nearly inaccessible village, the Hunzakuts have a record of genetic isolation quite unique in this world of mobile people.

It could be argued that the role of genetic factors in longevity might seem less important in the Caucasus. In Abkhasia, on the shores of the Black Sea, and adjoining the Caucasus mountains, one encounters many people over 100 who are not only Georgian (a provincial state in Russia) but also Russian Jewish, Armenian, and Turkish. The Caucasus is a land bridge that has been traveled for centuries by conquerors from both East and West, and its population can scarcely have maintained any significant degree of genetic isolation. At the same time, one invariably discovers that each of the centenarians in the area has parents or siblings who have similarly attained great age. The genetic aspect of longevity therefore remains a prime clue in solving the riddle of the aging process.

Longevity is clearly a multifactorial matter. First are the genetic factors, all who have studied longevity are convinced of their importance. It is generally accepted that the offspring of long-lived parents live longer than others, yet, a long life extends beyond the period of fertility, at least in women, so that length of life can have no direct biological evolutionary advantage. Living organisms seem to function much like wind-up clocks: their span ends when the initial endowment of energy is expended, just as a spring-powered clock stops when its energy source is depleted. L. V. Heilbrunn estimated in the 1900s that the heart of a mouse, which beats 520 to 780 times per minute, would contract 1.11 billion times during the 3.5 years of the mouse's normal life span. The heart of an elephant, which beats 25 to 28 times per minute, would have beaten 1.021 billion times during the elephant's normal life span of 70 years. The similarity between these two figures seems to suggest some initial equal potential that is gradually dissipated over the animal's life span.

Genes also influence longevity by predisposing the organism to specific fatal diseases. Alex comfort of University College, London, has pointed out the possible role of heterosis, or "Hybrid vigor." The cross-breeding of two inbred strains each with limited growth, size, resistance to disease, and longevity can improve in a wide variety of species, both plant and animal.

"One of the most remarkable features of the genetic code is that it is the same for all plants and animals that have been studied so far. These organisms include viruses, bacteria, yeasts, several plant species, and various animals including man; in other words, *the code is probably as old as life itself.*" (1973 Edition of the *World Book Encyclopedia.* Emphasis added.)

Next are the factors associated with nutrition. Since we are composed of what we eat and drink, it is not surprising that most people attribute longevity to dietary factors. The classic studies of Clive M. McCay of Cornell University in the 1930s showed that the life span of rats could be increased by the restriction of caloric intake in early life. Rats fed a diet otherwise balanced but deficient in calories showed delayed growth and maturation until the calorie intake was increased. At the same time their life span was extended by 40%.

The dietary aspects of longevity should largely be viewed as a suspension in the normal cell life span cycle – and not an actual increase in cellular productivity. No increase in the biological limit of an organism has been achieved through dietary control. Even if we could erase the cumulative wear and tear that affects the aging organism, it seems probable that the cells themselves are not programmed for perpetual activity. The actual difference in the life spans of various animals suggests some imposed limit. The changes that are observed in the tissues of animals that undergo metamorphosis – insects, for example – are indicative of the programmed extension of certain tissues. So is the failure of the human ovaries at menopause, long before other vital organs give out. Leonard Hayflick of the Stanford University school of medicine has reported that the cells in cultures of human embryonic connective tissues will divide some fifty times and then die. If growth is interrupted by plunging the culture into liquid nitrogen, the cells will resume growth on thawing – continue, that is, through the remainder of their fifty divisions and then die. Although one may question whether the cells are not affected by adverse environmental influences, Hayflick's studies support the theory of programmed death. Only cancer cells seem capable of eternal life in culture. One familiar line of cancer cells, the "He-La" strain, has often been the subject of cell-culture studies; it originated as a cervical cancer, twenty-one years ago, and has been maintained in culture ever since.

WHAT CONTROLS THE AGING PROCESS?

Collagen, the main protein of connective tissue, constitutes approximately 30% of all protein in the human body. With advancing age, collagen molecules show a spontaneous increase in cross-linking of their sub-unities, a process that increases their rigidity and reduces their solubility. Such a stiffening of this important structural component of our bodies might underly such classic features of aging as rigidity of blood vessels, resistance to blood flow, reduced delivery of blood through hardened arteries and, or a final consequence, the loss of cells and of cellular function. Other giant molecules, including the DNA molecule that stores the genetic information and the RNA molecule that reads out the stored message may also be subject to spontaneous cross-linking that would eventually prevent the normal self-renewal of tissues. It would appear that the DNA is genetically programmed to limit the RNA's cellular reproductive signal – and if this hypothesis is correct, and a chemical agent could be developed to "strip" this portion of the genetic code, or "short circuit" this encoded cellular death, then the deteriorating cellular evolution known as "aging" would be arrested, and a condition of perpetually suspended age would be achieved. Takaski Makinodan of the Baltimore City Hospital (among others) thinks that methods of rejuvenating the immune system are possible.

This essay is not meant to propound a particular theory, or extrapolate a particular area of research. The research into this phenomenon is far too limited to project a possible solution at this time. It remains to inform the reader of the possibility, if not practicality, of solving the riddle of man's hitherto most inevitable end, "death by old age," and utilize the solution to this biological problem for the obvious practical application of Terra II's voyage, but also for the continued life and, hence, biologically possible intellectual evolution of man here on Terra I.

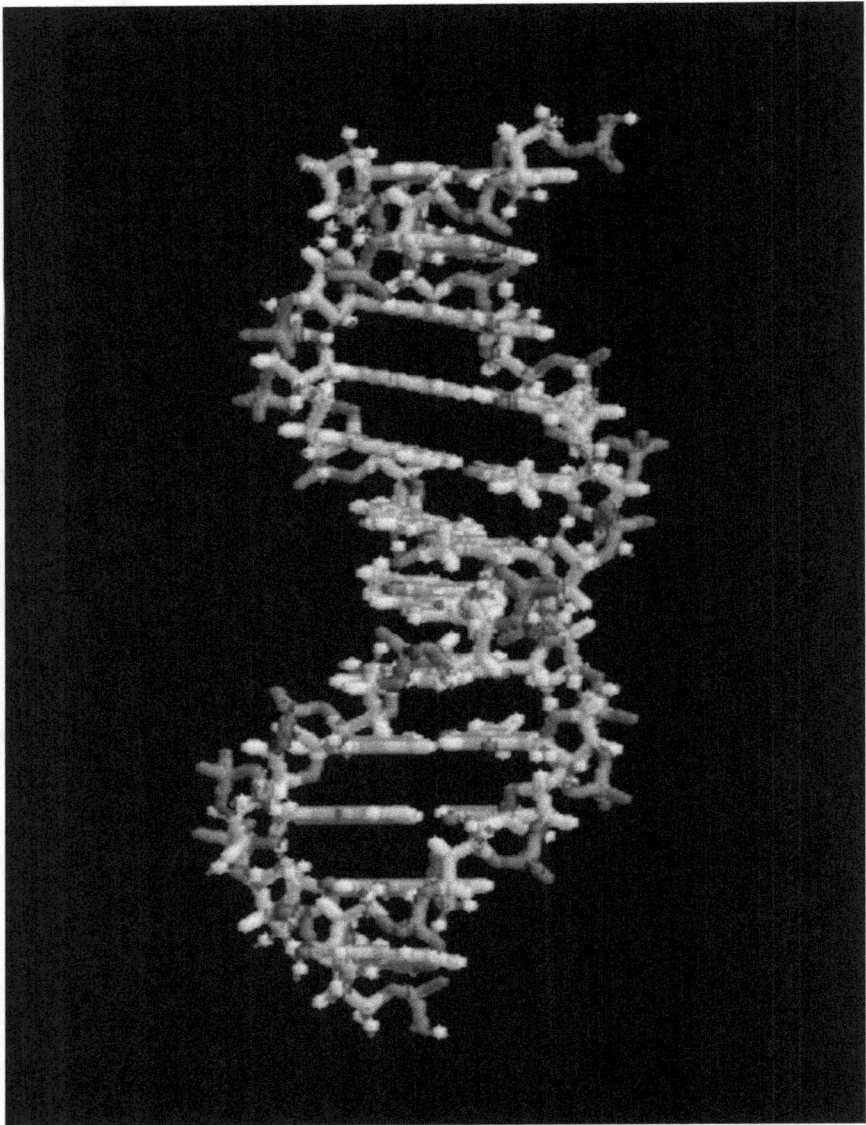

A model of a short segment of a DNA molecule. Each atom is represented directly. Actual DNA molecules may be thousands of times longer than the short segment shown here.

The DNA molecule is composed of four chemical ingredients:

Adenine (a), cytosine (c), guanine (g), and thymine (t). The strands themselves are made of sugars (s) and phosphates (p). Thus, a nucleoside is a combination of a base and a sugar, such as AS, while ASP is an example of a nucleoside phosphate. (a nucleoside phosphate with only one phosphorous group is called a nucleotide.) the sequence of bases specifies the genetic code by determining which proteins the cell will construct. Proteins, in turn, are long chains of amino acids. Recent evidence indicates that three nucleoside phosphates in the nucleic acid are required to specify each amino acid in the protein. The transcription sequence is this: DNA makes RNA; several kinds of RNA together make proteins, in particular, enzymes; and enzymes, by controlling the rates and varieties of chemical reaction in the cell, govern its metabolism. In this way, the nucleic acids actively control the form and functions of all cells. For several of these steps, protein intermediation is required.

Exact replication – the production of two identical DNA molecules from one – occurs because only certain combinations of bases can fit across the two strands. During DNA replication, the strands of the helix separate. The genetic material of all known organisms on earth is composed largely of DNA and RNA. These nucleic acids have coded into their structure the information which is reproductively transmitted from generation to generation. In addition, they have the capability for self-replication and mutation. DNA serves as a kind of molecular blueprint which controls metabolism, produces a replica of itself for the next generation to follow, and, through the centuries, gradually changes, or mutates, occasioning new forms of life.

The structure and function of DNA have been elucidated chiefly by the American molecular biologist James D. Watson of Harvard University, and the British molecular biologist Francis H. C. Crick of Cambridge University. DNA is a long molecule, comprising two molecular strands wound about each other in a coil or helix . . . During cell division, the strands separate, and each synthesizes a copy of the other, yielding two molecules of DNA where originally there was only one. This is the primary molecular reproductive event. The building blocks for this synthesis are called nucleoside phosphates. Much of the activity of the cell is devoted to constructing these building blocks from yet simpler molecules acquired from ingested food and joining them together to form nucleic acids. The nucleoside-phosphates are each composed of a sugar, a base, and some phosphates. A given nucleic acid molecule is generally composed of four kinds of nucleoside phosphates. Their sequencing along the chain is a kind of four-letter code that determines which sequences of amino acids, and therefore which proteins, a cell will make.

. . . For example, suppose an adenine-containing nucleoside phosphate is sitting bound in one strand. A variety of other nucleoside phosphates are available in the medium and occasionally come close enough for chemical bonding to occur. If a guanine-containing nucleoside phosphate is added, DNA synthesis will not proceed, because the guanine-adenine combination will be too large for the space available between the strands. And adenine-cytosine combination will not combine properly, nor will an adenine-adenine combination. Only an adenine-thymine combination fits between the strands. Elsewhere, a thymine-cytosine combination will be too small for the DNA double helix, and not reach between the strands. The replication of DNA occurs in large part because thymine (T) will only bind with adenine (A), and guanine (G) only with cytosine (C). Thus, once the sequence of bases along one strand is specified, the sequence along the other strand is determined. For example, if a section of one strand of a DNA molecule had the sequence of bases T C A G A G T G A C C G A T A T T C, we could immediately decide that the sequence of bases along the other strand must B E A G T C T C A C T G G C T A T A A G.

> – Sagan and Shklovskii, *Intelligent Life in the Universe*

THE CHEMICAL KEY OF IMMORTALITY

Dr. John Paul, director of the Cancer Research Institute, Glasgow, Scotland, reported to the 13th International Congress on Genetics, meeting in Berkeley, California, that, "we have determined that the unused sections of genetic material (DNA) are masked by proteins called histones." Figuring out how these "genetic sections are masked, could explain a host of human conditions," he said, "including cancer, aging, cell differentiation and evolution." Dr. Paul told the genetic congress that "It's conceivable that genes that are currently masked-off and unused, will someday be activated, and help create new strains of mankind."

Research into the aging process has proven that death is an encoded condition. Current research indicates that biologically speaking, there is no reason why cellular encoded "death" and the signals that trigger this "death" cannot be arrested. Immortality lies hidden behind a thin coating of chemicals (proteins called histones) that, when understood and "unmasked," will signal the end of programmed death in mankind.

Nearly all of what is known about DNA and genetic heredity as related to the process of aging has been discovered since 1970. Only two years after Watson-Crick determined the structure of DNA, Dr. Paul "discovers" that much of the code is "unused" and that "further evolutionary" development may be hidden in these masked areas. It is conservatively predicted that ten years of concerned research into the genetically controlled aging process will end the present cycle of programmed death. Project Starseed will inaugurate the first international research effort of an extensive nature into this phenomenon, and most people reading these words will live to see the time when "death caused by old age" will be an *unnatural* end to their existence. Again, it is predicted that *arresting* the aging process at the biological epitome of physical development will be realized within the next ten to fifteen years.

15. Each year added to the life span of the body must be accompanied by a multiplication of the range and speed of neural function

(This is the English translation of the 15th Transmission)

Implications of Immortality

Starseed seeks what the ancient religions defined as the goal of life. Immortality and celestial contact with the creator.

These goals will not be sought in metaphorical ritual but in actual practice.

There are two keys to immortality:

1. The "L" pill.

2. Time dilation created by close-to-speed-of-light acceleration

The "L" Pill

Recent discoveries in the fields of genetics and gerontology suggest that within a few years we shall understand the molecular mechanisms which produce aging. The process of senescence is caused by a chemical signal. When this signal is blocked the aging process can be stopped. It is quite likely that other chemical signals can stimulate the body to rejuvenate.

We have called this chemical intervention the "L" pill.

A moment's reflection will call to mind the many social and political problems that will be caused by the "L" pill. The death rate will fall. People will not die. The population might increase.

The "L" pill process has many implications. If people start living longer, they will be around in the distant future when restorative medicine increases its skill. Each decade of life increases the chance that medicine will cure the disease that is likely to kill you. You can live long enough until organ transplant has been perfected. You can live long enough until brain transplants make possible your turning in your old body for a new one. Etc.

The longevity business, once started, is endless in its potentiality.

We already know enough. If the governments of the world were to devote to longevity research one-fiftieth of the military budget we would have a practical "L" pill in two or three years.

Obviously, governments are not about to encourage an "L" pill. For all we know, the "L" pill may already be discovered, its existence censored because of the obvious political problems. There is some evidence that Russian scientists have carried longevity and brain transplant research far beyond that published in professional journals.

If the "L" pill discovery were to be announced the public reaction would be chaotic. Everyone will want it. The psychological implications strain the limits of our fantasy.

Who will get the "L" pill"? Will it be prescribed? Will it be restricted? By whom and why?

The situation will be very similar to that of LSD in America during the 1960s. LSD, when used knowledgeably, is less dangerous than the motor car or the airplane. The drug was, at the beginning, used only by specialized researchers who kept it from the public. Studies by Sidney Cohen on 25,000 cases showed that when used by persons who understood its application, LSD was not a dangerous experience. When banned by the American government and taken by ill-prepared persons in the context of a scare-campaign, LSD produced a casualty rate about a hundred times less than aspirin, but the psychological impact caused by millions of people changing their consciousness, developing new perspectives, brought down an avalanche of panic. All the established institutions reacted with enraged terror.

It is very likely that the same scenario will be played out with the "L" pill. Certainly, once the pill's safety and efficiency have been proved, the scientists themselves will be the first to take it. It must be remembered that the American government stopped certifying LSD research because in the early days the experimenters were taking the drug themselves and then reporting changes in their understanding of themselves, psychology, philosophy, etc.

People who take the "L" pill will start acting differently. They will be turning on, tuning in, and dropping out. They will quit their jobs, go back to school, divorce their spouses, return to their spouses, take off to spend five years on the beach, certainly religious attitudes will be changed, the very attitude towards life will be changed.

For these reasons we suspect that the government will not be in a hurry to license the "L" pill for general consumption.

The development and general use of the immortality pill will, however, have the highest priority on Terra II, this trip to the center of the galaxy may last for hundreds of years. The death rate and the birth rate can be carefully controlled. The optimal contract would be this: each crew member should have the option of living during the entire voyage and returning to the home planet younger.

TIME DILATION

While, according to the correct indications of their clocks, the crew has been traveling for about eleven years, to people on earth more than a thousand years will meanwhile have passed according to their likewise correct clocks. The two time measurements have experienced a dilation, which is completely analogous to the Lorentz contraction of distances.

If, during the journey to their star, it were possible for the crew to observe details upon the star they are aiming at, then prior to their start from the extraterrestrial station they will have observed the conditions on their star of a thousand years ago. As they are approaching the star with high speed, the last thousand years of the history of their star would be spreading out before their eyes in quick motion, and having arrived after eleven years, measured by their proper time, they would look upon what could be studied there as the present-day history of the star. Within eleven years, they would have witnessed a little more than the whole last thousand years of their destination star; that is almost a hundred times faster than an inhabitant of this star could have done. Meanwhile, during their journey to this star, a thousand years would have elapsed on the earth as well as the destination star.

– Eugen Sänger, *Beyond the Solar System*

Region of the galactic nucleus of M31. At relativistic
speeds a trip to the great galaxy M31 could be made in
approximately 28 years. During this voyage several million
years will have passed on earth.

Dr. Sänger, an Austrian, did fundamental work on rocket motors in the 1930s. He also did pioneering studies on what may be the ultimate propulsion system, the photon rocket, a device that would use the pressure of light itself to provide thrust, and might eventually take men to the stars. The paragraphs quoted above were taken from Dr. Sänger's testimony to the United States Congressional Committee on Science and Astronautics, printed in its report. *The Next Ten Years In Space.*

The longevity formula unfolds like an exponential equation. The current productive pace of science is such that each year one lives dramatically increases the probability of access to procedures which will extend one's life ten years or more.

Accident and sudden infection by alien bacteria become the only threats to a consciously controlled life span.

At this state of our knowledge it is safe to say that the most vexing, problematic aspects of longevity are not biological, but psychological. In the last quarter century we have seen the life span sharply extended. This gift has been accompanied by psycho-social disasters: boredom, nihilism, feelings of helplessness, social turmoil, generation conflict, increasing repression. We confront here a psycho-physical law: for each year that we extend the life span we must square the extension of consciousness. The body is the car and the nervous system the driver. It is meaningless and dangerous to improve the speed and mobility of the car unless we expand the consciousness of the driver. The extension of biology is arithmetical; the extension of neurologic must be exponential. People must have more and more to live for if they are to live more and more.

Now it becomes clear that Terra II is the ideal living laboratory for longevity. Terra II is truly a "time-ship."

Freedom is a key factor in the longevity formula. If life span is expanded without an exponential increase in the vistas of mobility and change, then the human race will become a technological anthill, a robotized, centralized Orwellian grim nightmare.

It is important to note that Terra II becomes an evolutionary experiment for Terra I. It will be possible, indeed, it will be necessary, for Terra II to initiate new ways of living, experiments in sociology, psychology, ethics, and neurology, which cannot be performed in Terra I societies. It is impossible, for example, to evolve new social forms on this planet. This was the lesson of the 1960s. Experimental mutants cannot occupy the same space as species dominants. Any visible changes in life style, sexual-social pattern, mode of consciousness, will be ruthlessly suppressed. The Catholic Pope opposes birth control and every police chief in the world itches to barber the long-hairs. Etc.

Starting four hundred years ago the Western Hemisphere became an evolutionary experiment for Europe. The "Melting Pot," they called it. Today the sophisticate European observes America as a fascinating and useful laboratory producing some things of value and many things to be avoided like the plague.

Terra II will serve this function for Terra I. Terra II is a seed-arrow launched into the future. Everyone left behind on Terra I will be stimulated and enlightened by developments aloft. We are consciously assembling the "elite" of humanity to explore the future. It is a democratic "elite." We expect that many other groups will assemble other time-ships to join the voyage into time. Humanity will be allowed to confront the implications of each star-bound philosophy. For example, some sort of conscious population control will be evolved. It would be most instructive for the Catholic Church to launch its own time-ship and thus test out its theories on birth control. Racial and religious conflicts which now involve the planet in spatial turmoil can be projected into time, each chauvinistic group can center its ambitions on its time-ship:

Most exciting of all the prospects (of interstellar contact) are the spiritual and philosophic enrichment to be gained from such exchanges. Our world is undergoing revolutionary changes. Our ancestors enjoyed a serenity denied to most of us. As we devastate our planet with industrialization, with highways, housing, and haste, the restoration of the soul becomes more and more inaccessible. Furthermore, through our material achievements, we are threatened by what the French call 'embourgeoisement' – domination of the world by bourgeois mediocrity, conformity and comfort-seeking.

The world desperately needs a global adventure to rekindle the flame that burned so intently during the Renaissance, when new worlds were being discovered on our own planet and in realms of science. Within a generation or less we will vicariously tread the moon and Mars, but the possibility of ultimately 'seeing' worlds in other solar systems, however remote, is an awesome prospect. 'The soul of man was made to walk the skies,' the English poet Edward Young wrote in the eighteenth century.

– Walter Sullivan, *We Are Not Alone*

EFFECTS OF TIME DILATION ON A
JOURNEY OF 4,000,000 LIGHT YEARS

SPERM-SHIP
DEPARTS
EARTH ORBIT

EARTH

DURING THE SPERM-SHIP'S JOURNEY
THE EARTH AGES 4,000,000 YEARS.

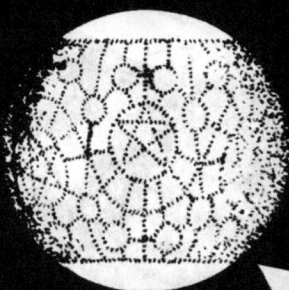

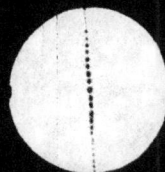

RETURN TO EARTH ORBIT
4,000,000 YEARS LATER

AS THE SPERM-SHIP ATTAINS NEAR-LIGHT VELOCITY, THE EFFECTS OF TIME DILATION REDUCE THE LONG JOURNEY TO THE NEAREST GALAXY AND BACK TO A SINGLE LIFE SPAN OF 60 YEARS FOR ALL ABOARD, WHILE THE EARTH ITSELF EXPERIENCES 4,000,000 YEARS.

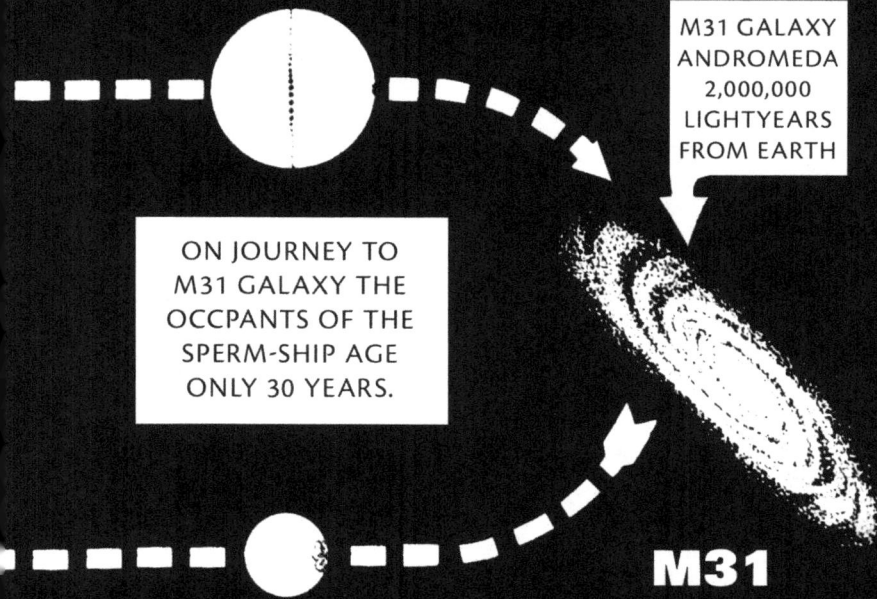

M31 GALAXY ANDROMEDA 2,000,000 LIGHTYEARS FROM EARTH

ON JOURNEY TO M31 GALAXY THE OCCPANTS OF THE SPERM-SHIP AGE ONLY 30 YEARS.

M31

ON THE RETURN JOURNEY THE OCCUPANTS OF THE SPERM-SHIP ALSO AGE ONLY 30 YEARS.

16. The sperm ship will have a spherical body and a propulsion tail.

(This is the English translation of the 16th Transmission)

The Architecture of Terra II

Terra II is a sphere. Diameter = two kilometers.

The "top" half of the sphere is sky, mountain, lake, earth. The sky-dome at its zenith is, of course, one kilometer high.

The bottom half of the ship (in this illustrative design) is divided into twenty-five levels, each forty meters high.

Levels 1 to 5, the top levels, might be residential. One thousand persons living at each level. Individual and communal buildings would take up approximately 100,000 square meters. Each of the five top levels will have approximately three million square meters of area. The remaining 2,900,000 square meters per level will be given to parks, lakes, outdoor recreational facilities, riding, paths, etc.

Level 6 could contain urban entertainment: shops, restaurants, theatres. These buildings take up ten percent of the area, the rest being devoted to gardens, lawns, walks.

Level 7 could contain 5,000 offices, classrooms, libraries, computer centers. The buildings would comprise less than ten percent of the space; the remainder devoted to open area, gardens.

Level 8 could house laboratories. Again, the buildings cover less than ten percent of the area.

Level 9 could house light manufacturing and assembly lines. Again, ninety percent of the area would be botanical.

Level 10 could contain hydroponic gardens.

Levels 11 and 12 could house heavy manufacture, smelting, forging. The area free for gardens and landscape may be less than fifty percent.

Levels 13 and 19 could be used for storage of supplies of all kinds. Approximately twenty percent of the area would be landscaped.

Levels 20 to 25 will hold raw materials.

About ninety percent of the volume of Terra II will be given over to "Nature." Sky, earth, water. There will be approximately four billion cubic meters within the sphere. Less than 200,000,000 cubic meters will be inhabited or used by the crew. There will be 40,000 cubic meters per capita for living and working space.

From these statistics it is obvious that Terra II will not be a sky-city, but rather a miniature earth. The atmosphere will be wilderness, rural, and suburban.

THE SHIP OF LIFE

Terra II is not a human venture. It is a message from all-life on the planet.

Every effort will be made to make the time-ship an ecological capsule in which the widest range of life forms on this planet will be represented.

At least ninety-five percent of the one hundred square kilometers of "floor space" will be luxuriant vegetation. Gardens, forests, lakes mountains. The human community will not depend on this land for food.

The optimal number of bird species will be included. The optimal number of wild and domestic animals. Reptiles. Amphibia. Insects, even at the risk of infestation. As many aspects of fish and marine creatures as possible. It would, of course, be hoped that dolphins will be included, both because of their intelligence and also for possible communicative purposes. It may be that the ship will encounter higher intelligences which exist in a marine environment where the dolphin's ability to communicate would be useful.

The exact population balance of non-human species will be decided by the ecological committees, but the general approach should be made clear. The time-ship Terra II should reek, bulge, swarm with life in all its myriad forms. Within the limits of inter-species public health, the ship should vibrate with flora and fauna. Just as no inch of our planet is without life, so with the seed-ship. Pets should be encouraged. Living quarters and offices wreathed in plants. Etc.

Terra II will sail through space, not a sterile, spic-and-span military craft, but an earthly vessel of life.

THE "EARTH" OF TERRA II

The "top" half of Terra II is wild terrain unencumbered by any human artifact. It will resemble the universe as seen from the Ptolemaic perspective.

The kilometer-high sky-dome can be transparent, providing a view of the starfields towards which the ship is floating. Or it can have an earth-simulated climate of daylight, clouds, artificial sun. The temperature will be varied to create a range of climates.

Two snow-peaked mountains (300 meters high) will dominate the landscape. Small rivers run to the central lake, a kilometer long and half a kilometer wide. One section of the terrain will be tropical, another temperate, another desert.

As many wild animals and birds as possible will inhabit the landscape. As many insect species as ecologically feasible will be brought aboard. Every attempt will be made to duplicate the ecology of the home planet.

Most of the landscape will be given over to forest and prairie.

The lake will be stocked with as many varieties of marine life as possible. Perhaps there will be a small salt-water lake.

The "top" level will be used as a wilderness area. There will be no residential or other buildings. Those who wish to camp out may do so.

Some of the land will be planted to agriculture; fruit trees, grains, vegetables. The food thus cultivated will not be used for stable nutrition, but for ceremonial and ritualistic purposes. The hydroponic gardens will be the standard source of food.

THE RESIDENTIAL LEVELS OF TERRA II

The first five subterranean levels could be residential and outdoor recreational.

Approximately one thousand persons would live at each level. Divided into four villages of 250 persons. The villages would take up less than five percent of the land on each level.

Each village would be distinguished by a cultural architectural style. The twenty cultures might be:

1. Japanese

2. Chinese

3. Southeast Asian

a. Polynesian

5. Indian

6. Middle-eastern

7. Egyptian

8. North African

9. East African

10. West African

11. Russian

12. Greek

13. Middle European

14. Italian – Spanish – Portuguese

15. French

16. Scandinavian

17. German – Swiss

18. English – Scottish – Irish

19. South American

20. North American

Each village will be a cultural unit with its own food center, community area, nursery, and small-child-care facilities.

The dwellings will be cottages and duplex apartments with garden areas and views of the galaxy without and the park areas within.

Each person over the age of puberty will have his-her own dwelling place. Approximately ten square meters in area. Couples who wish to live together will be able to double up in either of the two available "homes," alternate, or move into adjacent apartments which can convert into double homes of twenty square meters.

It is expected that most Terrans II will move from village to village. After a few years in the Japanese town, one may move to the French, etc.

Eventually Terrans II will have lived in all of the culture centers. It may turn out that ninety percent of the inhabitants of the Japanese village will be of non-Japanese origin.

The residential areas are for leisure. Except for child-care and food, all shopping and work will take place at other levels.

Ninety-five percent of the residential area will be park land, gardens, and outdoor recreational facilities. both individual and group sports. There will be a lake and riding trails at each level. Each area will have an outdoor natural theatre where the entire population can gather for entertainment and meetings.

THE URBAN LEVEL OF TERRA II

The sixth level will contain all the facilities of an urban-suburban shopping center. Shops. Restaurants. Dancing. Entertainment.

Shops. Everything is free. One's possessions will be limited only by storage space and the aesthetics of clutter. The shops, which operate like lending libraries, will contain clothing, personal ornamentation, aesthetic materials, personal electronic devices, recreational objects, specialty foods, etc.

Entertainment. A Tivoli garden complex. Musical and theatrical events. While every home will contain the most complex video equipment for personal education, recreation, and communication, the sixth level facilities would provide the place for communal gatherings, etc.

The seventh level, like the six above, will be given over to botanical and animal life, with five percent of the area built up. The buildings will contain the administrative offices, computers, electronic communication equipment, educational rooms, information storage.

This level is a major "nerve center" of the seed-ship.

A key factor in the mental development of the crew will involve neuro-technological imprinting and conditioning. The computers and electronics will become extensions of the nervous system. Every home will contain the linkages necessary to connect the individual with the memory and processing capacities of the computers. It will rarely be necessary for a person to leave home to communicate with anyone else on the ship or to perform cognitive activities. The seventh level simply provides the place for people to meet personally to plan, train, reprogram. And to conduct experiments in advanced neuro-electronic areas. Automation will be an added capacity which must never substitute for the stimulation and alchemy of personal contacts and direct manual-visual contact.

THE LABORATORIES OF TERRA II

The eighth level could contain the scientific laboratories – physics, chemistry, biology, medicine, etc.

The medical center could be located in this complex.

THE HYDROPONIC GARDENS

The ninth level contains the food supply for the company.

There can be as many as forty layers of gardens, each layer a meter high.

Food processing plants are also located at this level.

Nutritional and dietary research will take place here and in the eighth level laboratories.

While the diet is organic-vegetarian, food will be processed and packaged in every form known to the human race to delight the aesthetic taste.

RUSS 'SPACE MEN' GREW THEIR OWN FOOD

– New York

Four soviet researchers, in an unusual experiment looking toward prolonged space travel to distant planets, have spent six months in a closed biological environment, simulating the earth's biosphere.

The Siberian test project, reported by the soviet government newspaper *Izvestia*, is said to have demonstrated for the first time that man could live in an artificial atmosphere created by green algae, fast growing wheat and a variety of vegetables.

The plants were irrigated with sewage waste and with condensed moisture from the atmosphere. The researchers breathed the oxygen released by the plants, ate the vegetables, and used the wheat to bake their own bread during a simulated space trip in an underground chamber in the Siberian city of Krasnoyarsk.

A report of the man-plant interaction test was published October 27 in *Izvestia*.

The experiment, sponsored by the Krasnoyarsk Institute of Physics, is part of a continuing search for long flight life-support systems being conducted both in the Soviet Union and in the U.S. The aim has been to develop regenerative techniques that would eliminate the need for hauling vast amounts of oxygen, food and water on long space journeys.

Past tests have made use of physical and chemical devices to absorb the carbon dioxide exhaled by man and to recover oxygen. The soviet demonstration used green plants to convert the carbon dioxide into oxygen and food. The only thing coming in from the outside, according to the report, was an electric power supply.

The Siberian simulator was divided into four compartments of equal size, each with a volume of about 2600 cubic feet. This made the entire device somewhat smaller than the U.S. orbital space station Skylab, with an over-all volume of 12,000 cubic feet.

One compartment was used for living quarters and the three others for the cultivation of plants. The wheat field covered an area of 175 square feet, and the vegetable garden, with patches of potatoes, tomatoes, cucumbers, onions, carrots, radishes and lettuce, 35 square feet.

The chlorella algae were produced in three shallow-water tanks. Aside from their use as space food, these single-cell plants have been studied for about two decades as a potential source of high-grade protein and B-complex vitamins for overpopulated regions of the world.

To insure a continuous supply of wheat, the field was subdivided into plots on which a crop was planted (and harvested) at five-day intervals. The soviet report did not say how long the wheat took to mature under the powerful electric lamps, filled with xenon gas, which replaced sunlight. Fast-growing varieties require normally 100 to 120 days.

The experimenters tested different combinations of atmospheres during the simulated flight. In the first two months, they lived on oxygen produced by the wheat and the vegetables: in the second two-month period, by the chlorella and the wheat, and in the last stage, by chlorella and vegetables.

All three combinations were found acceptable by the humans, but the chlorella-vegetable combination, for reasons yet to be established, produced a wilting of vegetable crops.

New York Times, 1 November 1973

Most of the Terra II living quarters will be located about 200 meters above the upper "earth" level. In three bands against the inner hull (see illustration). The "earth" side of the living quarters will be faced in one-way glass, giving the occupants a view of the entire earth panorama making up the upper sphere. The one-way glass will blend with the sky-simulated hull (dome), making the residences invisible to those roaming the natural areas of the upper sphere. The outer walls of the living quarters will have "space view" panels. These will be tv screens, not windows. These same screens will allow wall-to-wall viewing on all intra-ship communications. Including closed circuit entertainment.

Elevators will be spaced every 30 meters and will join the upper living levels with all below-deck sections through their shafts between the ship's two hulls. Below the earth level, and all subsequent decks, transportation will be by: 1.) auto-walks (moving sidewalks, see cover art), 2.) com-trol (computer controlled) electric cars that surface in traffic circles when they reach their destination, but submerge between decks on tunnel tracks while en route (see cover art), 3.) paths (walking power).

The deck known as Eur-Asia might have a temperate climate and architecture simulating Europe and Asia. Because of total weather control, buildings will not have walls or other architectural fittings presently used to shield the occupants from the elements. All walls (where they exist) will be constructed only where noise screens are needed, or light filtration is important (i. e., film labs, theatre) or where atmospheric contagion is a problem (i. e., labs working with chemicals, virus, etc.). Light shielding for offices, etc., can be done by polarizing the air, dimming or blanking out light waves, or by glass, where absolutely necessary. Sonic screens will also be used to muffle noise.

The deck called Am-Af could have a tropical climate with architectural and landscape motifs from South America, Africa, Micronesia (South Pacific Islands), India, etc. Architectural representation will be largely suggestive. The cultural motifs will be carried out in furniture, landscaping, sculpture, symbolic cultural edifices, etc.

The subdeck C (and cold storage area) could have a 200-square-meter area of simulated arctic conditions. This area might include a frozen lake (for skating), moderate slopes for children's skiing, tobogganing, etc. This recreation area will only be a small part of C deck's arctic area; the rest will be a nature preserve with penguins, seals, and arctic birds and wildlife, etc.

The earth level deck could have a three-season weather variable: winter-(mild) fall combination, summer and spring. Clouds will be formed from moisture evaporating from foliage and the lakes, with artificially created cloud vapor, and rain as needed or desired. Lightening displays could be held in seasonal sequence. The earth level could have roaming wolves, mountain lions, snakes (some poisonous), bears, etc., and walkers might be issued "stun" sticks to ward off carnivorous predators. No killing of animals for sport will be acceptable under the "life value" concept. Children will be encouraged to develop an understanding of nature and wild creatures, to spend nights camped out, naked, without artificial aids, in the wild areas. Skill in communing with animals and living with the climates will be encouraged. A great event aboard the city-ship will be the planting of a new flower. People will breed hybrids in addition to maintaining earth "norm" strains. Cultural landscaping will be encouraged, i.e., Japanese gardens.

Dr. Robert D. Enzmann of Raytheon corporation has done a considerable amount of conceptual engineering of nuclear-pulse starships . . . Enzmann's initial design is probably one of the small, primitive, early-day starships.

Basically, it is a cylinder 300 feet in diameter and 1,000 feet long. A Saturn V without the Apollo LES escape tower would lie sideways across the inside of this cylinder. It contains nearly a half-million cubic feet of living space inside. New York City's Empire State Building would tuck neatly inside of it with just the top tower sticking out one end.

On the front end of this cylinder is a 'now ball' 1,000 feet in diameter made up of 12 million tons of frozen deuterium, the nuclear fuel for the eight Orion nuclear-pulse propulsion modules that will propel the ship up to 30 percent the speed of light (Mike 0.3).

The Enzmann starship isn't exactly just one ship. The cylindrical portion is made up of three identical cylindrical modules docked end to end. Each module is completely self-sufficient with its own auxiliary quarters, communication equipment, repair shops, storage holds, and EVA landing craft.

Each drum-like module is built upon a central core 50 feet in diameter and 300 feet long. Covering this backbone are eight decks of sub-modules each measuring 10 feet by 10 feet by 23 feet. These sub-modules are used as living quarters, storerooms, laboratories, and recreational areas by the human crew. Each of the drum-like modules has 700 of the smaller sub-modules.

The outer layer of sub-modules is used for communications equipment, EVA landing boat storage, observation equipment, laboratories, heat exchangers, and general storage. Their sheer mass helps provide radiation shielding for the living quarters further inward from the skin.

During acceleration by the nuclear-pulse units, the jolts of the explosions would be smoothed out by the shock-absorbers on each of the eight propulsion units. During the coasting portion of the journey, the ship would be spun-up around its longitudinal axis to provide artificial gravity for the people aboard.

Enzmann designed the starship to begin its voyage with a crew of 200 selected men and women carefully chosen for their skills, knowledge, and social stability . . .

The optimum starship population is 2,000 souls. This balances the designed closed-cycle ecology of the ship. Structuring the society of an Enzmann starship to maintain this balance as well as to preserve the original purpose of the flight pose some fascinating problems in applied social engineering. This is the really unknown variable in Project Star Flight. Perhaps we will have learned enough about it on earth beforehand: perhaps we will not, and Project Star Flight itself then becomes the laboratory.

But we wouldn't send just one starship to a promising target star. An interstellar expedition should be organized on the 'Columbus plan.' That is, it would be made up of a fleet of three to ten starships traveling together. Obviously, this has certain definite advantages. If something happens to one ship, its modules and sub-modules can be disassembled in flight and attached to another ship. All modules and sub-modules would be designed to be as interchangeable as bricks in a chimney. There is strength in numbers and survival in flexibility.

Thus, an interstellar expedition would be very much like a traveling space city with a population of up to 20,000 people.

– "A Program for Star Flight,"
by G. Harry Stine,
in *Analog*, October 1973.

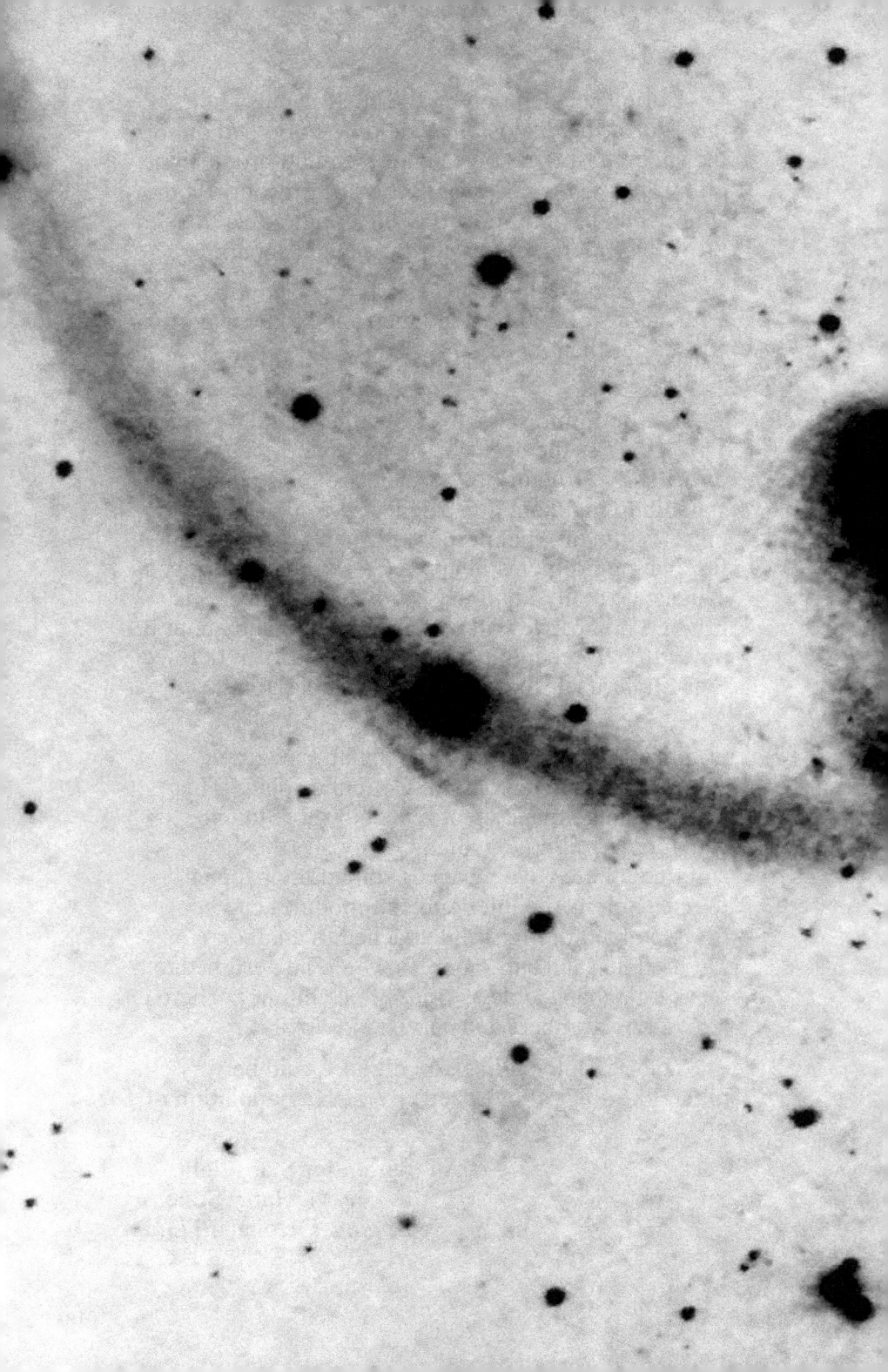

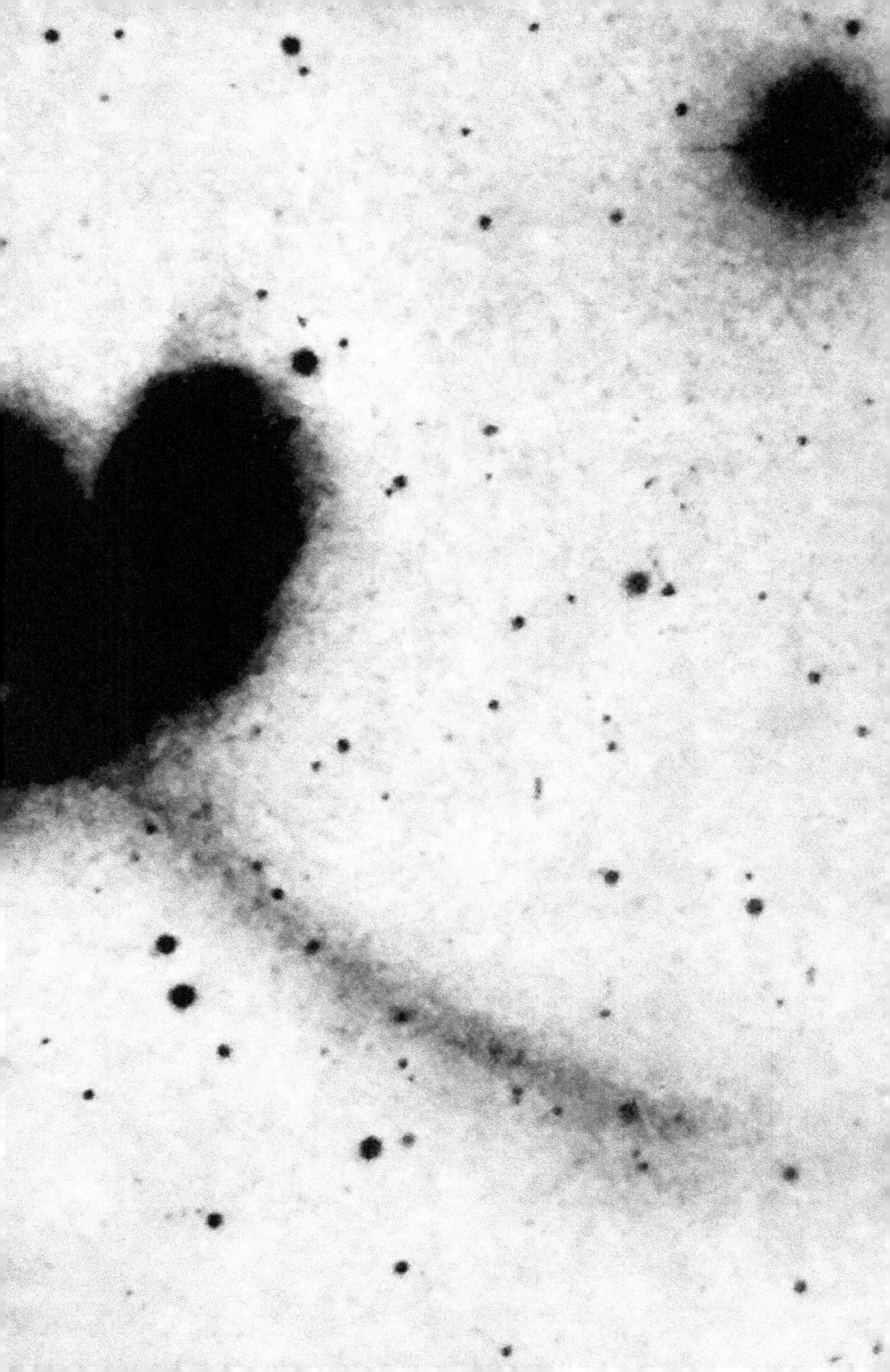

17. ORGANIC GROWTH PROCEEDS BY CELL DOUBLING, AT AN INCREASING RATE ACCORDING TO A GENETIC DESIGN.

(This is the English translation of the 17th Transmission)

WE ARE PROJECTING AN ENERGY WAVE

Terra II at first impression may appear grandiose and impractical.

Actually, Terra II, taken step-by-step, is alphabetically simple and practical.

The following time-table outlines how the Starseed project will unfold.

Starseed follows an organic plan. Embryology reminds us that the most complex structure can evolve in a short period of time by means of cell division following an overall design.

These lines are being written on October 22, 1973.

One year from today, October 22, 1974, Starseed and Terra II will be the most powerful concept in human consciousness unifying the planet with hope, excitement, and curiosity; over a thousand members of the pre-flight crew will be working full time on the project; millions of people in every country will be contributing their energies; the construction crew will be on site working on the Terra II replica.

The commonsense realism of this projection can best

be understood by comparison. Let the reader consider the specific, timetable plans of the other great national and international organizations. What do the United States, Russia, China have planned for the world by October 22, 1974? What does the united nations aspire to in the coming year?

It might be argued that retiring to a Japanese island and an Antarctic ice-flow constitutes an escape from the "real" world problems of war, plague, famine, pornography, overpopulation which concern the current world governments.

Our answer, of course, is that Terra II faces all the problems, physical and social, that now perplex Terra I. The five thousand crewmembers will have to develop, immediately, a perfect society. This assemblage of the smartest, most innovative people on earth will act as laboratory and inspiration for Terra I. Starseed is not only practical, it is necessary for the survival of the race.

The suggested timetable may seem rushed. It must be remembered that we are projecting an energy wave which will build momentum. As more and more persons contribute more and more of their force, the cumulative rate of growth will accelerate.

THE CALL

January : The call will be sounded in Kyoto, Cambridge, England,

Fountainbleu, Basel, etc. Groups all over the world will provide back-up publicity and stimulation.

THE DISSEMINATION OF THE INVITATION

January : The first 1000 copies of Terra II will be published in San Francisco California. Joanna will have a press conference at the U.N. building and present a copy to the Japanese delegation. Guanine will have a press conference in Tokyo and present copies to various Japanese officials and scientists. They will also announce the address of Starseed and the availability of the newsletter edited by Joe McKeon and Chip Roberts. Guanine and Joanna will continue all out publicity.

January : Guanine and Joanna start contacting the Nuclear Cell – respectable scientists, philosophers, writers, etc: Carl Sagan, Bernard Oliver, J. S. Shklovski, Arthur Clarke, John Lilly, Watson and Crick, Jean Paul Sartre. These people are not invited to join the crew, but consulted for assistance in arranging the conference and in contacting the appropriate Japanese savants. Michael Horowitz begins negotiations for publishing Terra II in several languages and for the manufacture and licensing of Starseed items designed by Dana Reemes: jewelry, clothing, toys. It may be necessary to earn over a million dollars in two months to sponsor the Terra II convocation. Terra II is translated into many languages.

January : Joanna, Guanine, and members of the Nuclear Cell go to Japan for publicity and meetings with the Japanese government. By this time the Japanese people will have realized the importance of their selection as the host country, the greatest honor that can be assigned to a nation. They will be asked to lease an uninhabited island to Starseed for a period of fifty years, and to host the convocation. In addition to the glory and the spiritual charge, Japanese industry will profit from being the most proximal suppliers to a 300 billion dollar construction project and being most closely involved in the scientific fall-out. During the early months of 1974 we expect that Starseed will be the center of national discussion in Japan. The enthusiasm, hospitality, creativity, and competence of the Japanese will be a major intensification factor.

February : The Japanese delegate to the United Nations announces to the General Assembly that Starseed Convocations will be held in Japan during March, April and May, 1974. Every country in the world is invited to send the names of savants to the selection committee. Scientific and professional groups throughout the world are similarly invited to submit names. The invitation also is extended to all individuals in the world who believe they can contribute to one of the forty-nine plexes (committees). Letters are addressed to Starseed, Japan.

Selection of those who will attend the Japanese Convocation will be made by a personal-contact network operating via telephonic hookup. Let us assume that astronomer Carl Sagan becomes an enthusiastic participant. He is given a telephone credit card and proceeds to phone the scientists whom he knows might be interested. They will have been sent a copy of Terra II before the call. Several telephonic conversations may be necessary to exchange all the information necessary for decision. He recruits seven scientists who simply agree to come to the Japanese Convocation. Each of these is requested to search through his colleagues and professional acquaintances for seven other authorities on any of the 49 scientific topics. If six other "seed" scientists like Sagan perform the same chain-letter personal linkage, over 300 invitees will have been selected – an average of seven per category. This process of selecting pre-flight Convocation experts must not be confused with crew selection which will take place over the subsequent ten-plus years. The criteria for crew selection will emphasize adaptability, emotional flexibility, intelligence, creativity, charm, beauty, genetic endowment, wealth, etc. Selection of pre-flight Convocation experts will be on the more limited basis of professional prestige, competence in field, open-ness of mind to new ideas, sympathy with extra-terrestrial exploration. Wise, planful minds, each of the 300+ experts thus self-selected will be asked to choose seven additional invitees from lists submitted by governments, professional societies and from individuals. The invitation process will go on for two and a half months; the great majority of invitees will have been selected in the first month of the process.

TABLE I
Scientific Plexes: Japanese Convocation

1. Mathematics

2. Physics

3. Computer science

a. Electronics-electrical engineering

5. Astronomy-Astro-Physics

6. Astronautics

7. Agriculture-Landscaping

8. Chemistry

9. Chemical Engineering

10. Medicine

11. Neurology-Pharmacology

12. Bio-Physicis. Bio-chemistry

13. Public Health-Ecology

14. Exo-Biology

15. Physiology

16. Genetics

17. Geology

18. Botany

19. Zoology

20. Architecture

21. Manufacturing, Tool Design

22. (To be added by Seed Plexes)

23. (To be added by Seed Plexes)

24. (To be added by Seed Plexes)

25. (To be added by Seed Plexes)

The February announcement by the Japanese U.N. delegate will be followed up by world-wide publicity by the Starseed staff. We wish every person on the planet to be informed about the project and invited to join either as an applicant for the Japanese Convocation or as local agent. Applicants will be requested to write a letter to Starseed, Japan describing their interests, aspirations and potential contributions. It is expected that hundreds of thousands of letters will be received.

During the period December 28, 1973 to March 1, 1974, Starseed elements will travel around the world generating public interest, contacting potential invitees. Each day will provide newsworthy items of interest in preparation for the Japanese Convocation which will be the most widely covered assemblage in world history.

THE SCIENTIFIC CONVOCATION

Two convocations will be held: a scientific convocation lasting the month of March and a cultural convocation the month of April.

On March 1, 1974, there will assemble in Japan between 1000 and 1400 scientists and engineers. The meetings will be held in a resort hotel if possible away from urban complications.

The participants will select one of twenty-five plexes (committees), as listed in Table I. There will be approximately fifty persons in each plexus.

At the initial meeting, a few papers will be presented summarizing the evidence and theory in that field which is relevant to Terra II. The plexus will then divide into sub-groups which take up aspects of the general field. The division into task-groups will be prepared in advance by planning groups, open to amendment by convokees. Each morning the plexus will meet as a whole to summarize the work of sub-groups and then divide into smaller units to allow individual participation. Each plexus is free to reorganize itself and reschedule its activities. Every third or fourth day the entire Convocation will meet to receive summaries of the productions of the plexes.

The Convocation will last for one month. It is to be expected that some participants will not be able, or interested in staying, for the entire month. During the proceedings new members may be added. Each plexus will have democratic control of its own activity.

All proceedings of the convocation will be televised and taped. It is expected that a large press delegation will be on hand and daily programs will be broadcast. Television and radio rights will be sold to national and international networks to pay for the expenses of the Convocation. It is likely that a daily one hour summary of the convocation will become the most popular global television program, linking the planet in fascinated observation.

Technical staffs will be on hand to help each plexus: computers, audio-visual demonstrations, mechanical mock-ups, artists, photographers, etc.

The convocation will dazzle, entertain, educate, and inspire the world. Since there will be representatives from every nation, a high degree of local interest is assured. Nigeria, by way of illustration, might have five delegates, five observers, and a press delegation of five. The television audience of Nigeria will be treated to a nightly one hour program summarizing the most newsworthy events of the preceding day, illustrated with diagrams, animated drawings, narrated and explained by Nigerian experts. The result of this signaling will lift humanity's eyes and visions to the stars, reinforce local-cultural pride, and help weave each isolated community into the fabric of a global aspiration.

The functions of the scientific convocation are:

1. Production of facts and ideas about technical aspects of the voyage

2. Global education and consciousness-raising publicity

3. Selection of pre-crew who will continue to work full-time on the project

4. Formation of the organization which will accomplish the preflight activities

Table II

Cultural Plexes: Japanese Convocation

1. Political science

2. Economics (incl. fund-raising)

3. Anthropology

a. Sociology, Social-Psychology

5. Psychology-Psychiatry

6. Systems Analysis

7. Ethology

8. Public Administration

9. Linguistics

10. Education

11. Library Science, Literature, Archivism

12. Communications

13. Culinary Art

14. Music

15. Visual Arts

16. Physical Culture

17. Performing Arts, Entertainment

18. Film-Television

19. History

20. Philosophy-Theology

21. To be added by seed plexes

22. To be added by seed plexes

23. To be added by seed plexes

2a. To be added by seed plexes

THE CULTURAL CONVOCATION

April 1, 1974

On April 1, 1974 approximately 1200 persons will assemble in the same location for the Cultural Convocation. To this group will be added approximately 125 participants of the Scientific Convocation who will have quit or received leave from their jobs and stay on for the second Convocation.

The participants will assign themselves to one of twenty-four plexes which cover the social aspects of Terra II. There will be approximately fifty-five persons in each plexus. The list of cultural topics is presented in Table II.

The plexes will meet, divide, re-assemble and

communicate as did the scientists. Lessons learned about organization from the first Convocation will be applied.

Television coverage of the Cultural Convocation should be even more fascinating to the global audience in that the topics will cover basic issues of social life, practical political science, communication, sexual mores, family life, education, child-growth, group and individual psychology, boredom, psychopathology, religion, philosophy, ethics, aesthetics architecture, control of violence and crime, etc.

It should be clear that the Japanese Convocations will become a global university, a planetary educational experience, a world-wide consciousness-raising event. Every issue raised by participants and every solution proposed will have direct relevance for the stay-at-home audience. Humanity will be privileged to witness its wisest and most endowed planning a new society which will be based on all available human knowledge and guided by a collective vision of harmonious community. The televised Convocations will start global discussion of all the ancient problems in a new context in which no one can be threatened or imposed upon.

The functions of the cultural convocation are:

1. Production of facts and ideas about the communal and personal aspects of the voyage

2. Global education and consciousness-raising

3. Selection of pre-crew

4. Organization for pre-flight

THE PRE-FLIGHT ORGANIZATION

May 1, 1974

Around 300 of the convocation participants will remain in Japan during the month of May to initiate the pre-flight organization. From this point until departure, the energies of the company will be structured into two geographically separate projects:

Earth Island

Antarctica

Earth island, located on an island leased from Japan, is the site of the launching. Here will be designed and tested all of the constructional elements of the time ship. Here will be assembled all the supplies and equipment for the voyage. And from here everything and everyone for the voyage will be ferried by rocket to the orbital parking place (around the moon) where the ship will be assembled. By July 1, 1974, the first wave of the construction crew will be on the island setting up temporary living quarters and beginning the work of preparing the island for inhabitation. A harbor and airport facilities will be the first stage. Design teams located in Japan, on the island and around the world will be drawing up plans for the manufacturing and storage facilities.

The Antarctica Project is responsible for the construction of a replica of the time ship in the icy wastes of the polar cap. The building of the replica under difficult climatic conditions will simulate the building of the time ship in moon orbit. The gradual construction, deck by deck, of the replica and the gradual peopling of the craft will be a simulated test-run. As crew-members are selected they will be ferried to Antarctica and live in the enclosed conditions of flight. During the pre-flight years the crew will live and work exactly as they will in space. The hydroponic gardens, the communications systems, the living arrangements, the social forms will be experimentally tested. The crew members will be able to test themselves. The results of this experimentation will be important in preparing for flight. The results of these experiments will not involve any final decisions because the flight conditions will be continually adapting and evolving.

Terra II will carry the equipment and supplies necessary to reconstruct any aspect of its own design. And Terra II will carry a crew selected primarily on their ability to adapt to and evolve into new social forms. Antarctica will be in close televised contact with earth island. Each laboratory and office of Antarctica will be hooked up to its counterpart on earth island so that conferences and continual-simultaneous communication will be possible. The first survey crew will reach the Antarctic region by September 1974.

During the month of May 1974 the 300 participants in the third Japanese Convocation will set up the preliminary structure for the two pre-flight cities and for the world-wide Starseed network.

THE STARSEED NETWORK

The function of Starseed is to broadcast.

The electric message has been received. Leave the earth-womb and live in the universe!

This message is neurologically irresistible. It generates a spark in everyone who receives it. It is the ancient signal that every living entity has been prepared to receive – awakening the dormant instinctual cellular readiness.

The message must be transmitted. This "must" is not a moral imperative but a lawful energy process. Light "must" shine. Life must propagate.

This signal will be received by every human being. "LISTEN! WE ARE GOING TO HEAVEN! WE ARE GOING TO THE STARS!"

The most "primitive" native in New Guinea will automatically understand. *A Message has come from the stars telling us to come.*

The Japanese Convocation will be the most widely broadcast event in human history.

The apollo moon landing, by comparison, increased helplessness and decreased hope. The world was forced into a passive spectatorship watching American soldiers unfurl flags, recite the King James Bible, unveil a plaque inscribed with the name of Nixon, hit a golf-ball. Etc. How could the school-boy in Nigeria relate to these chauvinistic Houston pranks? By contrast everyone on the planet will be invited to write Starseed, Japan. A correspondence network will emerge. Each correspondent will be sent *The Starseed Journal*, a monthly newsletter, to be translated into their native tongue. And a pamphlet of suggestions for further involvement. There is something that everyone on earth can do to facilitate Terra II. Translate and distribute the books. Form Starseed groups in the local community. Since Terra II involves the preservation and glorification of all life forms and all aspects of human life, every person will be invited, as minimum participation, to select and collect aspects of local culture to be included in the new world. Every book written in every language will be microfilmed and added to the Terra II archives. Every aspect of Nigerian culture can be filmed, taped or miniaturized for transmission. Terra II is literally a sperm-ship carrying in its archives the miniaturized electric-seeds of every aspect of evolution. Everyone can help construct their own molecular link in the great DNA chain of Starseed. We shall be able to reproduce all forms of terrestrial life and all aspects of human history for the entertainment of extraterrestrial intelligence.

Time relativity is the haunting theme. By the time Terra II reaches the galactic network thousands of years will have passed on earth. Almost nothing will be left of "modern" culture. The DNA archives of the sperm ship may be the only extant evidence of our history. By the time Terra II reaches the center of the universe (perhaps a few decades on board) billions of years will have transpired in this solar system; the planet earth will no longer exist. Only Terra II floating through galactic time will carry the record of life on earth.

When these factual possibilities are realized every human will be motivated to contribute an energy-memory. The Republican Woman's Club of Pasadena (a very conservative city in California, U.S.A.) may conduct bridge parties to raise money to microfilm anti-vivisection propaganda which, in turn, might be the first earth-data we would show to a Higher Intelligence possessing furry coats and four gravity-grabbing pawlike appendages. Etc. The John Birch society or the Communist Manifesto might just preserve our reputation for sanity in some galactic ant hill. The glory of Terra II is that it assumes that every Terra I event was a necessary step in our evolution.

From this position of totally accepting liberation, we can only be reverently grateful to those who struggled to get us to this free, joyous launching-spot. Nixon-Agnew and the Israeli-Arab war may have been the last grand embryonic contortions to show us that it is time to be born and get about our galactic business. The agony of Nixon can thus be understood and sympathetically cherished. (Mr. Nixon, we hope and pray that you read these words which offer you the only way that you will obtain what you seek – the approval and love of all humanity.)

By the summer of 1974 dozens of Starseed cells will have emerged in every city on earth. By October 22, 1974 it is possible that more than half the world population will be involved, however minimally, in contributing to Terra II. There is nothing in this plan that excludes any human being or any previous human aspiration.

At this point a new "political" perspective comes into focus. The plan projects this awful possibility. Starseed may become the most "powerful" organization in the world, linking up a majority of human beings in one "cause." There is a certain spiritual (i.e., neurological) truth to this suspicion. Starseed will become the most influential "idea" in the world. We can accept this responsibility of world "leadership" because Starseed is totally pure at the material level. We want nothing from this planet except to leave it and to carry with us whatever any individual or group wishes to be preserved. John Wayne's Oscar award will be microfilmed and brought to heaven. The Swiss school-boy's collection of bubble-gum cards will, if he wishes, be preserved.

We confront here the politics of consciousness. We have no wish to destroy, change, interfere with any human activity. We assume that on the day of "lift-off" Irish-protestants and Irish-Catholics will still be killing each other in Belfast, we shall gladly microfilm their manifestos and take the opinions of both sides to the center of the galaxy. Etc.

Every other vision, religious program, movement in human history has implied that people have to change what they are doing to conform to the new utopian crusade. The only philosophy which can excite the interest and cooperation of persons from every country is one that truly *renounces* all worldly ambitions.

Starseed and Terra II, by definition, truly renounce all worldly ambition.

Consider the reality picture from our perspective. This planet is a grain of dust in the galactic brain-cloud. We have no more interest in "taking over" planet earth than we have in becoming wardens of the particular prison in which we now find ourselves. We can honestly repeat the classic truisms of conventional religion. The competitive affairs of this earth have little concern for us. Terra II is a state of mind. A level of consciousness. We are truly detached from mundane desires. We can sincerely say that our ambitions are not of this world.

CONTINGENCY PLANS FOR CONVOCATION AND SELECTION

The plans just outlined for international conferences may seem unduly optimistic, given the current state of national rivalry and social inertia.

The plan will succeed on a grandiose and accelerated scale simply because there is no other alternative. Starseed offers to each individual and each country precisely what everyone wants: global cooperation for immortality and conscious evolution. Starseed implicitly asks the question: if you do not want this convocation to meet in Japan, what *do* you want to happen in the spring of 1974? What is your alternative proposal? Starseed takes from no one. It gives to all individuals and countries exactly what they want from it.

Skepticism, inertia, opposition will be used to generate more energy, to clarify, test and strengthen. If we cannot deal with Terra I obstacles we are obviously not ready to conduct dialogue with Higher Intelligence.

The time table here outlined assumes very rapid growth and cooperation from a small but bold minority of recognized scientists. The danger of this plan resides in the swiftness of its growth. Too many people, too soon, with too-high expectations may lead to disillusionment. Etc.

Opposition to the plan and a slowing of its growth will be most useful in testing and developing the strength of the vision.

Here are some contingency options:

If the Japanese people do not accept the Starseed

invitation to lead the world to the stars we shall find another small country. At the present time Otto Palme and Bruno Kreisky are the most intelligent, liberated, cultured, far-seeing leaders in the world. Sweden and Austria will be invited to host the Convocation. If political changes prevent this alternative, the Convocation would be held in the United States or Canada – larger, more heterogenous locales. In any case the world will be watching and learning from our efforts.

We expect the Convocation to be organized around a small number of respectable, prestigious scientists. History does not teach us to be sanguine about enlisting the aid of "established" scientists to support an evolutionary development. They are notoriously conservative. If recognized scientists fail to come forward we shall shift our invitation to younger, not-yet-famous scientists. There are certainly more than enough young instructors and graduate students in every academic department and research laboratory who know as much as their distinguished colleagues, but who would be open to the new vision. The Starseed strategy is to approach the older and more conservative first, to invite them affectionately and respectfully into the future. To avoid the repetition of a "Generation-gap conflict." If the older and politically powerful experts refuse, the problem is simplified. Fifty young astrophysicists will be more effective than fifty Nobel laureates. Etc.

If Starseed should fail to enlist the support of enough young professionals, the invitations will go to undergraduate colleges and to high school students. There can be no doubt that there are fifty such young students of astronomy, chemistry, sociology, etc. throughout the world who would respond with enthusiasm.

Terra II will have a strong appeal for thoughtful young

people and every effort will be made to prevent older generations from being cut off from the adventure.

Another set of contingency options must be prepared in the opposite direction: enthusiastic acceptance by national governments. The Starseed vision backed by respectable scientific support will excite tremendous popular support. Nations with strong central governments which control public opinion may be indifferent to the plan, probably participating by sending delegate-spies as they do to other international gatherings.

Nations with a competitive democratic system of electoral politics may respond with more energy. The party in power may see Starseed as a popular issue and commit itself strongly to the Starseed plan. The political advantages of integrating national energy to Terra II are most seductive. For the idealistic young: an adventurous vision. For the bored aging: longevity, immortality, a renewal of interest. For the uneducated poor: a circus. For the educated: an intellectual challenge. For the industrialists: new business. For the pacifist: a plan for world harmony. For the military: the possibility of an enormous N.A.S.A.-type agency. For the politicians: a bandwagon movement to bureaucratize. Etc.

It is quite likely that one or more countries will attempt to nationalize Starseed. It is possible that president Nixon of the United States might see Starseed as a way of unifying his country. If half of the current military budget were diverted to Terra II construction, America could build and launch the time ship without raising taxes. The efficiency of the present N.A.S.A. organization plus the enormous scientific resources of the American government would make Terra II a simple matter.

To have the massive resources of the richest government in the world thrown into the project would be a tempting alternative. It is, however, basic to the Starseed vision that it be international and completely independent of any government. Political vagaries must be avoided. The Nixon administration might back the project. The next administration might not.

Starseed will accept the help of any individual, group, nation. It would be advantageous for Terra II scientists to collaborate with American and Russian and Chinese scientists; perhaps to perform joint experiments at Cape Canaveral and at Soviet and Chinese and French astronautical centers. But the territorial integrity-autonomy of Earth Island and Antarctica and the self-control of the operation must be maintained by Starseed.

While Starseed respects all governments it depends on all people.

Alistair Cameron, the astrophysicist, in the introduction to his anthology on inter-stellar communication, describes the possibility of life in other worlds as 'currently the greatest question in scientific philosophy.' Already, he says, we are admitting that there be millions of societies more advanced than ourselves in our galaxy alone. If we can now take the next step and communicate with some of these societies, then we can expect to obtain an enormous enrichment of all phases of our sciences and arts. Perhaps we shall also receive valuable lessons in the technique of stable world government.'

– Walter Sullivan, *We Are Not Alone*

Terra II ... A Way Out

18. OTHER SPERM SHIPS WILL FOLLOW.

(This is the English translation of the 18th Transmission)

OTHER TERRAS TOO

Public opinion surveys suggest that Starseed will rapidly become the most popular idea in human history.

Since the voyage to the stars is inevitable, the only question is when? When the signal is sent out that the trip has begun, the idea will grow exponentially and irresistibly.

Starseed is based on reverence for all life and for every human emotion and action. Even the great villains of past and present history are seen to have played their part in providing the broad insight that we now possess. They show us how we do not want to go.

While the majority of human beings will become involved in the voyage, only five to seven thousand will make the trip. This means that only one out of five hundred thousand humans can join the crew. The model for Terra II is DNA.

There will be many alternate spiritual models and mythic visions which could guide and pattern such an expedition.

It cannot be denied that Terra II is an elite of elites.

The Terra II elitism is personal. No group, class, creed or race will be excluded.

It is inevitable and necessary that other ships and other crews make the trip. National, religious, and racial rivalries being as they are, it is probable that some countries will consider sending their own ship to the stars. America, Japan, Russia, China are economically powerful enough. Catholicism and Islam.

Our attitude towards such "rivals" is totally affectionate. Nothing about the Starseed flight is secret or menacing. All of the scientific and technical discoveries will be publicly shared. Terra II will carry no arms. (This statement must be clarified. The Terra II shops and factories can produce any piece of equipment known to humanity. It would be possible for Terra II engineers to construct the most advanced weapons during flight. Consciousness of this genetic potentiality is part of the trip. Just as the DNA code possesses the ability to create armored dinosaurs, sabre-tooth tigers and B-52 pilots, so does the sperm ship carry within it the seeds for offensive and defensive weaponry.)

In selecting the crew, human and computer evaluation techniques will scan the recruited population for all genetic characteristics, including competitive aggressiveness. Some "warrior" seed will be included. This is to say that the crew will not be made up exclusively of passive, pacifistic gentle people. Terra II is not an escape in any sense. Every life potentiality will be aboard. Thus there is no guarantee that competitive-aggressive types will not be able to "take over" the ship and convert it into an intergalactic, bomb-laden warship. Constant openness and vigilance of consciousness is our only hope.

Nor do we have any assurance that a competitive group might not attempt to sabotage the enterprise during pre-flight stages. It will be very easy to spread rumors about the activities of Earth Island and Antarctica. Every hidden desire and twisted lust within the human psyche will inevitably be projected upon Starseed. The hungry of the world will accuse us of gluttony. The poor of exploitation. The ugly will charge us with erotic/narcissism. The sexually frustrated will accuse us of conducting orgies. The angry will suspect us of plotting against them. Etc.

Total openness and the electric precision of our wisdom are the only defenses. Starseed attempts to assemble the most intelligent and beautiful and courageous. Starseed claims to follow a pre-destined genetic blueprint. If, with these energies, we cannot charge, enlighten, inspire, uplevel the embryonic forms of our species, then we are not ready to sail for the stars.

While we must anticipate every possibility, we cannot fear failure or dedicate energy to defensive postures.

Terra II is the noblest, highest, most advanced concept that the human species has produced. If we fail, we fail in an all-out effort to contact the Higher Intelligence. Our failure is the most valuable experiment ever conducted. Future generations will study our design and learn from our experiences. Thus, we cannot fail. We cannot err. Everything we do from this moment on is celestial signal.

THIS BOOK IS A DNA SIGNAL

Let us frivolously imagine that UFO's which recently flew over Mississippi are inhabited by superior beings who are trying to communicate with humanity. They sit around scratching their heads. What to try next. We have given them the message in seed, in mathematical formulae, in epic psychodramas. To attract their attention we have appeared to them in the shape of common tableware, have assumed the form of little green creatures, etc. We have doused them with cosmic waves, sun-flares. We have jiggled their chemistry and pharmacology. Jammed their neural circuits with electro-magnetic impulses. Shocked them. Dosed them with LSD. Now let's send them a comet and Terra II.

The book you are holding in your hand is the greatest evolutionary artifact ever produced. It is a DNA signal from the Higher Intelligence. No matter the immediate fate of Terra II. The fact that you are reading these lines means that the transmission has reached those whom it was designed to reach. No matter what *you* do about it, this message is being transmitted through you and will influence the next phase of evolution. Your reaction is datum for the greatest experiment ever designed by humanity. Terra II is now permanently imprinted on *your* consciousness. From this moment on you will move towards Terra II or in competition with it. In either event you are contributing your weight to the evolution of consciousness.

We assume that other time-ships will be designed and launched. Starseed will cooperate in every way. It would be a charming comfort and exciting asset to have companion ships as we spin through galaxies billions of miles from earth. All the scientific and exploratory activities would be doubled in scope. Survival indices would be benefited if we could call upon the help of another time ship. Diversity of seed and structure is the basic lesson of DNA. A synthesis and exchange between two different time ships increases beauty and survival efficiency. We welcome competition.

And recognize the dangers. Suppose a rival sect or creed creates a time ship prepared to destroy heretics? Etc. Terra I history is not reassuring about the harmonious motivation of human collectives. Prospective visions of space-piracy, galactic-buccaneering. What is to prevent another time ship from capturing Terra II, robbing its riches and raping its women? Etc.

These stellar projections of the destructive aspects of Terra I life will not materialize. We grant the possibility that boredom, claustrophobia, flight psychopathology could covert a time ship into a seething craft of negativity. But we assume a metamorphosis, a neurological mutation. Caterpillars lumber around like tanks, devouring and destroying all organic material that falls in their grasp. Butterflies soar in erotic beauty, with no capacities to inflict injury. With open meadows of the galaxy ahead of us, with a universe of discovery beckoning, with the lessons of earth-territory-dominance behind us, linked into an electro-neural network, totally and instantaneously interdependent for survival and entertainment, the destructive competitions of embryonic earth *will be transcended.*

We expect that within a few years every human being on earth will be involved in a time ship project. Today, almost every person directly or indirectly contributes to the arms race. In the near future the same energy will be devoted to time ships. The first Terra II will cost 300 billion dollars. But later versions will be less expensive. If mass-production techniques were employed it would be possible for an industrial country to produce time ships the way we now produce nuclear subs or aircraft carriers. Scores of star-ships will be launched every year. We now realize that the discovery and exploration of the new world after 1492 was a prelude for our discovery of, and entrance into, galactic society.

Terra II ... A Way Out

19. THE FOUR-POLAR MAGNETIC ARRANGEMENT OF THE NERVOUS SYSTEMS IS THE INSTRUMENT OF RECEPTION AND TRANSMISSION.

(This is the English translation of the 19th Transmission)

TECHNICAL NOTES ON THE RECEPTION OF THE STARSEED TRANSMISSION

For those who are curious about or technically interested in the method used to receive the Starseed transmission . . .

This message was received, transformed, and transmitted by the classic method of neurological alchemy! Monastic isolation, intense sexual magnetism, continued exposure to solar radiation. Etc.

(Alchemy is the basic science from which most human knowledge has originated. Mathematics and astronomy are historically based on the study of energy cycles and rhythms, e.g. Astrology. Chemistry and pharmacological medicine are based on alchemy. Modern psychology-psychiatry-positivist philosophy are based on ancient energy-flow concepts of nature. Pythagoras. Heraclitus. Democritus. Hermetics. Paracelsus. Dr. Dee.)

We assume that the evolutionary process and the energy laws on this planet are the same as those found throughout the galaxy and the known universe.

We assume that the evolving human nervous system can receive, analyze-synthesize, and transmit an ever-increasing spectrum of the biochemical and electro-magnetic energy fields at play on this planet.

Reception of energy signals is facilitated by certain instrumental arrangements. The microscope, the telescope, electro-magnetic coils, radio- telescopes, for example.

We speak here of the process of capturing, re-arranging-recharging, and emitting energy.

Reception of signals is useless to the evolutionary process unless the synthesis and storage is followed by eventual emission.

The instrumental unit by means of which human beings receive, synthesize, and transmit signals is the nervous system.

The nervous system is designed and built by the genetic code.

The instrumental unit by means of which DNA receives, synthesizes, and transmits energy is the amino acid molecule quartet.

Four amino-acid molecules form the basic structure of life.

These four molecules, thymine, adenine, cytosine, guanine, come in polar pairs.

The electro-magnetic chemical field generated by these four units sets up a pattern upon which all life is based.

The human equivalent of this energy-conducting model is obtained by arranging four nervous systems in magnetic pattern. Two women. Two men. Two male-female pairs.

JANE DEE, JOHN DEE, JOAN KELLEY AND EDWARD KELLEY

An account of this neuro-magnetic technique is contained in the biography of Dr. John Dee, sage counselor to Queen Elizabeth of England, Dee was born of Celtic descent on July 13, 1527. After graduating from Cambridge he joined the founding faculty of Trinity College, just established by Henry VIII. Dee taught Greek and Philosophy "and enlivened his teaching by producing the comedy of *Peace* by Aristophanes. *In one of the scenes of this play, a man carrying a basket of provisions is transported to heaven on the back of a huge beetle,* and Dee managed the stage-effects so cunningly that many vain reports spread abroad of the means how that was effected; the consensus of opinion was that Dee had invoked the aid of magic, and he thus earned a reputation that many of his later doings served to confirm."

In 1553 Dee, with an annual pension from Edward VI, became a rector, later declining a professorship in Mathematics at Oxford. One of Dee's scientific interests was the prediction of cycles of human behavior based on chronologically recorded time sequences. He submitted an horoscope reading to Elizabeth Tudor in which he predicted that she would soon become Queen. He was thereupon arrested by federal agents sent by the understandably concerned Queen Mary. His apartment was searched and sealed and he was tried for treason. During his subsequent imprisonment he was interviewed in depth by government theologians and philosophic agents who found his predictions to be based on accurate astronomy and mathematics.

Upon the accession of Queen Elizabeth in 1558 he was restored to favor and performed services of council and prognostication for the ruler. After a lengthy trip through Europe engaged in information-gathering for the crown, Dee was given a royal benefice which allowed him to buy a country house where he initiated extensive studies in psychology, neurology, and metaphysics. Dee's central concern was the development of instruments and techniques for receiving and decoding electromagnetic impulses. These psycho-physical experiments using optical devices, hypnosis, yoga, meditation-concentration, and psychoanalytic free-association to detect unconscious impulses were considered by the superstitious to be forms of magic.

The queen consulted him on regular occasions and he performed the role of advisor, psychotherapist, and guru for the sovereign whose long, wise, and enlightened reign provides testimony to the skill and good sense of Dr. Dee.

It is of interest that in the Elizabethan court, renowned for its intrigues, Dee managed to maintain a secure position, most probably because of his detachment and serene indifference to material politics. *At one point Dr. Dee was called to Windsor Castle to explain the significance of a comet which had appeared in the sky.* Dee's quiet, scientific approach quieted the hysteria which invariably attended the comet phenomenon.

By 1580 Dee was totally involved in his experiments which won the respect of the intellectual community. His pharmacological and optical-diffraction research convinced him that he was receiving messages from Higher Intelligences, who in the pre-scientific language of the time were called angels and spirits. Dee was described as "tall and slender, with a very fair, clear and sanguine complexion and a long beard as white as milk. He wore a gown like an artist, with hanging sleeves and a slit. He was a great peace-maker; if any of his friends quarreled he would never be content until he had reconciled them."

Some of Dee's attempts to receive impulse-messages involved the manipulation of light rays, either through a crystal ball or a polished disk. Concentrated attention on light rays reflected and diffused seemed to put observers into a trance-like state during which visions and images were "hallucinated." The person in trance was called a *scryer*. Dee recorded the scryer's reports on paper and attempted to construct a new language to transmit the findings.

In 1582 Dee began working with a scryer named Edward Kelley, an ex-convict whose ears had been cropped for felonious theft and fraud. The results were so successful that Kelley joined Dee as regular scryer, the ex-convict and his wife moving into the house of Dee and his wife.

For the next eleven years Dee, Kelley and their wives worked together in telepathic and psycho-physical experimentation, in England, in Poland, Prague, and Bohemia – at that time centers of psychology, yoga, pharmacology and consciousness expansion. The fame of Dee and Kelley continually attracted the jealous enmity of the Catholic Inquisition which was fanatically opposed to any psychological experimentation. Giordano Bruno, during his visit to England, had become friendly with Dee and this contributed to the subversive reputation of the good Doctor and his charming, ex-convict friend.

At one point Dee and Kelley were hounded by Inquisition agents from one country to another on the continent, always suspect by the police and feared by the common people who had been led to think them magicians. The rumor that Dee and Kelley and their wives had formed a group marriage was true. According to Dee's records: "on Sunday, the third of May, Anno 1587, I, John Dee, Edward Kelley, and our two wives (Jane Dee and Joan Kelley), covenanted with god, and subscribed the same, for indissoluble and inviolable unities, charity, and friendship keeping between us four; and all things between us common, as god by sundry means willed us to do."

The exact nature of the research findings and techniques developed by this quartet have been lost in the perverse mists of controversy. It is clear that the bold experimenters were enthusiastic about their results. However, the combination of optical experimentation, drugs, yogic practice, novel linguistic, group marriage was obviously intolerable to the religious orthodoxy and the Dee-Kelley data and methodology were ignored in the storm of scandal and libel that swirled around them.

It is significant that Dee never lost the confidence of the shrewd and realistic Queen, who, as late as 1589, gave him a research grant of 100 marks, an annual subsidy of £200, and the wardenship of Manchester College.

Conservative historians, influenced by ecclesiastical prejudice, have made Kelley into a scoundrel and Dr. Dee into a lovable dupe. Modern scholarship, however, including that of Carl Jung, has established that alchemy was not a gullible or fraudulent attempt to convert baser metals to gold, but rather a consistent school of psychology which specialized in consciousness expansion and widened spectra of reception. And the concept of male-female sexual magnetism united in physical and neurological love as basic unit for attainment of higher levels of consciousness and for reception of messagery beyond the spectrum of the "normal" mind has become a convention among those familiar with Tantric Hinduism, Tantric Buddhism, Nyingmapa Tantrism, etc.

Quite by unpremeditated *coincidence* this same quadraphonic structure was used for reception of the Starseed transmission.

It happened like this.

Folsom is the maximum security prison in the state of California. Here are sent the most dangerous, intractable, unrepentant, incorrigible inmates in America's largest state prison system. Constructed in the early 1900s, the buildings are heavy granite. Five tiers of dark stone and metal. There are no amenities or cosmetic flourishes.

In architecture and behavioral routine the situation is Benedictine. The prisoner inhabits a narrow 4 ½ -by-10-foot single cell. Each cell door is locked by a long bar which runs the length of the tier. Six times a day a guard pulls a lever which slides the bar-tab away from the door for a period of ten seconds so that the prisoner can quickly enter or leave his cell. In clement weather the prisoners are allowed to visit the yard, a baseball diamond surrounded by asphalt walks. Once on the yard the prisoner cannot return to his cell until the scheduled "lock-up". Entering the prison buildings from the yard the prisoner walks directly to his cell to be present when the long bar slides open for the ten seconds. There is a line up for access to the dining hall three times a day. Unlock for the yard. Lock-up to cell. Saturday afternoon a movie in the dining hall. Week after week the same narrow routine of cell solitude and yard time. Time and space being totally controlled the inmate is thus totally free.

In the summer of 1973 two prisoners began a series of daily discussions walking back and forth across the grassy portion of the main-line yard.

The two prisoners were unusual, even by the grim standards of Folsom.

Lynne Wayne Benner, 32, had spent seven and one half years in solitary confinement (the hole) for anarchic militancy and non-violent dissent. During this period of almost total isolation he would go weeks at a time speaking less than a dozen words. Most of this time of diamond-pure Trappist isolation was devoted to reading philosophy, designing mechanical inventions, six of which were patented, working out detailed plans for utopian social systems and writing effective alternatives for prisons. Etc.

Four months before the transmissions he had begun a friendship with a beautiful woman who held a responsible job on the leading newspaper in a nearby metropolitan area. One day a week the prisoner and the woman would spend six hours of intense, loving communication in the institution visiting room. The remaining one hundred and sixty-two hours of the week the prisoner, almost totally free from distraction, thought about her, their future together, etc. If the mind can be freed from despairing bitterness and negative thinking, then the maximum security prison is an ideal place for meditation, centering of thought, and telepathic transmission and reception.

The second prisoner had just spent four months in an isolation cell with no interpersonal communication except for weekly visits with his mate, a beautiful woman who was totally dedicated to broadcasting his message.

To summarize: both prisoners were spending six hours a week in highly charged sexual interactions and all-out communications with their lovers and 162 hours a week in undistracted serenity. Neither of the prisoners nor the women were involved in negative thinking and all four were obsessed with hopeful mentation, joyful anticipation, and telepathic communication.

In August 1973 the two prisoners began walking together in the yard. A hypnotic-yogic routine. One hundred paces across the grass strip and both wheels, turning, with the formation precision of flocking birds. It so happened that during the summer of 1973 there was not one cloudy day in the area of Folsom, California. The two men spent five hours a day walking in solar radiation, barefoot, dressed in white athletic shorts.

The prison yard is probably the closest modern equivalent of the Athenian open-air academy. The caste-class society has been so arranged that there is nothing for the long term inmate to do except think with his mind and exercise his body. Outside the walls millions of taxpayers labor to support the institute of higher learning. In the kitchens food is being prepared. Maintenance crews work to provide shelter. The librarian is ready to order through inter-library-systems any book that might be required. In five gun towers marksmen guards lean down with rifles cradled in their arms to insure instant law and order. The prison population is an enormous neural network whose dendrites watch every television show, scan every periodical and signal any item of interest. Among the prisoners there is every profession and talent available for immediate voluntary consultation. Folsom prison is thus an electronic Cluny, a space-age monastic center preserving the light during the dark periods of convulsive despair which rack the "outside world."

In this climate, conversation quickly moves to ultimates. The cosmological questions. Where are we going? Where do you want to go?

One day in the sun Lynn Wayne Benner confides his vision. To leave the planet earth by space-ship in expectation of contacting higher intelligence; to utilize current findings in neurology and genetics to suspend the aging process.

"You want to use current scientific advances to attain immortality and go to heaven in a space ship!"

Benner nodded.

"My friend. That is the first intelligent, sensible thing that any man has ever said to me. Practically, how can it be done?"

The two men are in the prison yard walking directly into the sun. Then they turn and walk one hundred paces away from the sun. The hot, clear rays are on their faces and then on their backs.

Benner begins to talk about the astronomical, astronautical problems involved in inter-stellar flight. He has spent ten years researching the subject. Then he summarizes the current evidence on the bio-chemistry of aging, Russian work on brain-transplants. Statistics. Names. Logical alternatives. Dimensions. Prices. The man is totally inspired.

After five hours of absorbing conversation the prisoners return to their cells, at three-thirty in the afternoon. For the next nineteen hours they are free to meditate, review, read, dream.

They concentrate on two things. Their mates. And the way out. The plan for future life with their mates.

Both men have gone beyond middle class fears and aspirations. In their past both men had won middle-class success, fathered children, attained material affluence and mutated beyond in some inchoate, driven search for intensity, change, adventure, expanded meaning. In the darkest holes of society they had snipped the last strands of attachment to the rewards and goals of the middle-class and its government. Without hesitation both had risked life, reputation, security, sanity in the quest for transcendence. Sunk to the bottom of the Black Hole and reemerged in the blue-sky sunlight shining, glowing, radiating.

There is nothing novel or strange about this mutation. Every prophetic philosopher in history has walked willingly into this dark pit. It is the classic process. The familiar, classic all-out test. There is no way to avoid it. Please note that there was no sacrifice for a "cause," no killing for Christ or Allah or Mao. This is the yoga of pure and total freedom. The ordeal of all-out courage, after which comes the revelation.

Each morning the un-lock bar slides and the prisoners walk to the yard. The sun just won't stop shining. Both are bursting with the fruits of the nineteen-hour meditation. Benner is usually scryer. Each night the recorder types the receptions which are read in the sun. The scientific books needed to advance and guide the visions show up synchronistically. Another prisoner lays Sullivan's *We Are Not Alone* on the cell bars. The librarian orders Shklovskii and Sagan.

The process is automatic and unself-conscious. This is to say, the prisoners did not plan to set up this reception structure. Only later did it become apparent that towards the end of the solar scanning, around two thirty in the afternoon, would the Recorder be able to focus on a particular question and the scryer suddenly begin to sweep perspective up to an interstellar reference point from which the planet earth was seen in its galactic space-time reference.

The concept of Terra I as embryonic and Terra II as post-natal provided the Archimedes leverage from which one could literally move the earth. Etc.

And each Monday the two beautiful mates would come and the star-shine, sun-love connections would be made.

Put Yourself in the Position of Higher Intelligence

The signals were received in the form of neural patterns, images. Benner talking from the perspective of a galactic observer. Each night the recorder would type a few pages summarizing and illustrating the day's transmission.

The signals came in no logical order. The Recorder would pose a question and the scryer might answer with Terra I logic or fact . . . But sometimes, unpredictably, there would begin singing a revelation from the interstellar perspective.

It became clear that the Galactic Intelligence is here on

the planet continually transmitting. The barrier has been the inability of the human nervous system to open up and receive the signal. From the perspective of Terra II the prisoners anticipated Watergate, Agnew, the comet Kohoutek, the outburst of brutality, Chile, Sinai, and, of course, the current wave of U.F.O. Reports.

The trick for understanding U.F.O.s is to assume, temporarily, that Higher Intelligence does exist and is broadcasting throughout the galactic network. Then, put yourself in the place of the Higher Intelligence. Do not view It as alien. Try to become a collaborator, a receiver. Throughout the world when strange signals are received people automatically call the police. The police automatically reach for their guns. The expectation is that the emissaries from galactic center are going to be brought to the station to be fingerprinted and mugged. Journalists reach for their cameras. "Take us to your leader" does not mean take us to President Nixon.

Courage and hope are apparently key factors. No one on the planet really wanted to get in touch with Higher Intelligence more than the two prisoners and their mates.

"There is no one left to talk to, except Higher Intelligence."

"We can never be free to live with our beloveds until the planet is free."

The problem in interstellar communication is attention. Higher Intelligence beams electro-magnetic signals on the planet. Everyone is too busy with embryonic competitions and worries to listen.

The signal.

The "instruction" comes. Put the Starseed transmission in the simplest A.B.C. primer form that will be attended to and understood by embryonic minds.

The Recorder sits in his cell and types the Starseed transmission. A message from inter-stellar intelligence to planet earth. It is not to be distributed in typical Terra I form. It is not to be submitted to an editor. The copyright is not to be discussed by the publisher's lawyers. Did Moses sign a contract for the Ten Commandments? Did Christ get royalties for the sermon on the mount? Think of the two prisoners as interstellar energies attempting to contact humanity.

The Starseed Transmission is based on images, revelations, visions received by a quadro-magnetic human unit. The images were then translated, transformed into English. Unavoidably the style is that of the Recorder because the transmission was not received in the English language.

THIS BOOK IS A SIGNAL FROM HIGHER INTELLIGENCE

The transmitting quartet believe that they have been used as relay agents. There is not elitist implication. We believe that every human being is designed to be a relay station, receiving, transforming and passing on the galactic message. We say simply: this is the message we have received from Higher Intelligence. We ask humbly: what message have you received?

As scientists we describe as carefully as possible the experimental conditions. Two men totally detached from the pressures of earthly life, totally in love with their mates. Two beautiful, poised, "successful" women totally dedicated to their men. The experimental approach is important. Courage, hope, love.

As scientists we inquire of our fellow terrestrials: what methods are you using to contact Higher Intelligence?

This book is the signal which we received. We invite

everyone to join this circuit. We remain open to your signal and your invitation.

We have all come together and we are all going together.

Terra II ... A Way Out

20. THE KEY TO INTERSTELLAR TRAVEL WILL BE FOUND IN THE MANIPULATION OF GRAVITY.

(This is the English translation of the 20th Transmission)

ASTEROID PROSPECTING

The time-ship and most of its technical equipment will be manufactured in moon city factories.

The moon is carboniferous and low on metals. Deep-drill mining might provide the necessary metals to construct Terra II, but a more economical way of obtaining the raw materials might be to capture an asteroid.

One possibility would be sometime in 1975 to launch an interplanetary craft for the asteroid belt between the orbits of Mars and Jupiter. The ship would carry a crew of up to nine persons: pilot, co -pilot, navigator, metallurgists, mineralogists, and mining experts. The asteroid selected would be determined to be rich in iron and nickel and approximately one mile in diameter. The asteroid Hermes (diameter one mile) passed within 400,000 miles of earth in 1937, a distance only twice that of our moon.

The asteroid selected would be blasted out of orbit by atomic power and rocketed to lunar orbit and then dropped to the surface of the moon.

One of the main functions of island city is to serve as launch site for lunar-bound flights.

Chemically powered rocket-ferries will transport supplies to earth orbit stations. These vehicles, stubby-winged rockets, will discharge their cargoes in earth orbit and return to earth via retro-rockets. They will have jet engines, aeronautical flight controls and retractable landing gear. They will return to airstrips on Earth Island. It is expected that these ferries will be used for over thirty round-trips before need of overhaul.

Lunar shuttles will transport supplies and passengers from earth orbit to moon orbit.

These vehicles will be assembled in space and will never land. Arriving in lunar orbit they will dock at lunar orbit stations and transfer cargo to lunar landing modules.

Cargo will be transported in containers.

The first cargoes dropped on the moon surface will contain living quarters, pressurized domes and necessary life-support equipment.

One of the first steps will be to set up a radio observatory on the dark side of the moon where, free from the electronic noise of earth and (periodically) from solar radiation, eavesdropping on galactic messagery will be initiated.

The "earth-egg" is protected from radiation by the Van Allen belt, which operates like an amniotic membrane, screening out not only lethal or genetically disturbing radiation but also radio messages which might be sent from galactic sources. The enormous racket of solar radiation also drowns out incoming signals.

This protective envelope around the earth-egg serves exactly the same function as the embryonic membranes around the fetus. Adults know that it is pointless to attempt to communicate with a fetus by post-natal methods. We cannot give the fetus a book to read, nor teach it to respond to symbolic language. We have to wait until the fetus emerges from the womb to initiate post-natal conversation.

In precisely the same fashion, it is impossible for humanity to learn galactic languages and communicate with Higher Intelligence until we leave the membrane protection of the planet.

The Starseed Transmission suggests that the evolutionary purpose of the moon is to act as "telephone-receiver" and exit station for planetary life. The chances of receiving messages from galactic sources are multiplied a hundredfold by the use of radio-telescopes located on the dark side of the moon. The cone of receptivity created by the moon-shadow projected out in space like a beam will scan a wide band of interstellar sky.

It is quite possible that eavesdropping scans will pick up intelligent messagery. This will encourage the enterprise and suggest one feasible direction of flight. It is possible that the radio-telescope will pick up probe signals attempting to initiate conversation. If these travel at the speed of light there will not be time to send back responses. Terra II will have left the solar system by the time the signals are received. It is possible that such scanning messages would contain compressed volumes of linguistic and scientific data which could be decoded and provide technical aids to the project.

It is also possible that faster-than-light messagery will be received which might contain instructions on how to construct and utilize this valuable asset.

CONSTRUCTION OF MOON CITY

The main task of moon city is to construct Terra II.

Domes, blasted-out caverns, tunnels will provide living

and working atmosphere. In *2001, A Space Odyssey*, Arthur C. Clarke presents his version of a moon station:

Now there were new, strange stirrings on and below its surface, for here man was establishing his first permanent bridgehead on the moon. Clavius Base could, in an emergency, be entirely self-supporting. All the necessities of life were produced from the local rocks, after they had been crushed, heated, and chemically processed. Hydrogen, oxygen, carbon, nitrogen, phosphorous – all these, and most of the other elements, could be found inside the moon, if one knew where to look for them.

The Base was a closed system, like a tiny working model of Earth itself, recycling all the chemicals of life. The atmosphere was purified in a vast "hothouse" – a large, circular room buried just below the lunar surface. Under blazing lamps by night, and filtered sunlight by day, acres of stubby green plants grew in a warm, moist atmosphere. They were special mutations, designed for the express purpose of replenishing the air with oxygen, and providing food as a by-product.

More food was produced by chemical processing systems and algae culture. Although the green scum circulating through yards of transparent plastic tubes would scarcely have appealed to a gourmet, the biochemists could convert it into chops and steaks only an expert could distinguish from the real thing.

The eleven hundred men and six hundred women who made up the personnel of the base were all highly trained scientists or technicians, carefully selected before they had left Earth. Though lunar living was now virtually free from the hardships, disadvantages, and occasional dangers of the early days, it was still psychologically demanding, and not recommended for anyone suffering from claustrophobia. Since it was expensive and time-consuming to cut a large underground base out of solid rock or compacted lava, the standard one-man "living module" was a room only about six feet wide, ten feet long, and eight feet high.

Each room was attractively furnished and looked very much like a good motel suite, with convertible sofa, TV, small hi-fi set, and vision-phone. Moreover, by a simple trick of interior decoration, the one unbroken wall could be converted by the flip of a switch into a convincing terrestrial landscape. There was a choice of eight views. This touch of luxury was typical of the base, though it was sometimes hard to explain its necessity to the folk back on earth. Every man and woman in Clavius had cost a hundred thousand dollars in training and transport and housing; it was worth a little extra to maintain their peace of mind. This was not art for art's sake, but art for the sake of sanity.

It is necessary to build and maintain the replica city in the Antarctic because it will be ten years before moon city is ready. Replica-city will test each technical, biological and social form. The results of Antarctic experimentation will determine lunar design. Eventually all of the crew of Terra II will arrive in Moon City.

Some time late in the year 1999 Terra II will leave its moon orbit and begin Terra's grandest voyage. This metal globe will contain the representative seed pods of all earth life, the creative "flowers" of mankind. This relatively insignificant sphere will be the biological message of earth, flying on the wings of time out toward the center of the universe. The trip will be made, its duration and final destination depending solely on the time speed of the voyage. In that context let us consider a few of the means of propulsion to be studied and selected by the technical plexus during the international conferences beginning in Japan in January 1974.

Any discussion of the technical aspects of the Terra II flight must be qualified by probability that great advances in energy-engineering will occur in the next twenty-five years.

The Starseed Transmission specifically suggests that Gravitational Manipulation can be used to facilitate the voyage.

Freeman J. Dyson in his article "Gravitational Machines" written over a decade ago, said, "The difficulty in building machines to harness the energy of the gravitational field is entirely one of scale."

Gravity is the basic force of the universe. Gravity holds the atomic structure together. Gravity fixes the orbits of galaxies. Gravity holds us to earth. Even light waves respond to the pull of gravity.

When we split the atom we are manipulating gravity.

When the principles and practices of gravity manipulation are mastered we shall be able to rearrange atomic elements. We shall be able to project force fields which will change every aspect of our lives. With Gravity Manipulation it will not be necessary to build Terra II on the moon. It will be possible to lift off the ship, indeed, lift off the Earth Island, if we so wish, and spin it through deep space to our destination in the stars.

With Gravity Manipulation all structural concepts change. There will be no need for beams, floors, walls. Everything will be supported by controlled force fields. There will be no need for fuel and propulsion engines. Control of magnetic forces of attraction and repulsion will enable Terra II to use any material structure to "Push away from." The mass of earth, which now holds us captive, can become the repulsive force against which we can push. The mass of sun, dust-clouds in space, the galaxy itself, become attractive or repulsive centers. Everything is fuel for movement towards or away from.

Gravity is the fabric of the universe and we shall whirl through the galaxies on waves of basic energy.

Traveling at 10,000 G's of acceleration there will be no sensation of movement. Every molecule in the body will be accelerating equally.

The discovery of gravity manipulation will also allow the sperm ship to be constructed in a fraction of the projected time.

The technical aspects of this transmission must be tentative because of the rapid acceleration of our knowledge.

The design and architecture of the star-ship presented herein is suggestive. And based on current theories and facts about propulsion and construction engineering.

The evolution of our methods and conceptions will proceed at a spectacular rate once the work is begun.

The eventual goal is to develop Gravity Manipulation techniques based on the manipulation of atomic particles.

The term Gravity Manipulation embraces a wide spectrum of engineering methods.

Muscular propulsion, walking and running, is the basic form of gravity manipulation.

Mechanical engineering is the second evolutionary phase.

Electrical methods of design, communication, and propulsion become the next phase – far superior in speed and scope to mechanical. Nuclear forms of propulsion and energy-generation in spite of the enormous power produced are limited by the structure of the machines to which they are harnessed.

The strategy of flight evolution will probably require that the takeoff and first few years employ nuclear-powered propulsion, and that the architecture of the star-ship use conventional gravity-control devices, beams, floors, walls, storage containers, wires, etc.

Even before flight the radio-telescopes on the silent side of the moon will be seeking contact with Higher Intelligence. Certainly as soon as the Sperm Ship leaves the solar system the chances of being discovered by Galactic Intelligence will be greatly increased. Experimentation on improved methods of propulsion and gravity manipulation will continue aboard Terra II. After brief years of flight it is expected that contact will be made with civilizations which possess ultra-light drive, atomic force-field mastery of gravitation.

With this knowledge it will be possible to reconstruct the Star-ship entirely. The use of force-fields will eliminate the need for all structures as we now know them. Walls, floors, machines, tools, furniture clothing – will all be replaced by force-fields.

It is probable that within a few years the need for the Sperm-ship as vehicle will be transcended. Using gravitational force-field technology any couple or group of T.T.'s could spin off in their own bubbles. Terra II will become a floating home-base, the master DNA molecule of ex-terrestrial life.

The most recent news from the laboratories tells us of the existence of thychons, anti-energy units whose lower limit of travel is the speed of light. It seems almost certain that within a short time Terra II will contact intelligences who understand and control these energies. As we approach the speed of light – from either side of the energy spectrum – we are approaching the eternity-unity~simultaneity point.

Within a lifetime of the elect readers of these lines, it will be possible to move immediately to any point in the farthest galaxy. Advanced methods will make possible instantaneous re-corporalization of the body, if we still use bodies, and instantaneous reconstitution of the force-field patterns that make up our consciousness.

The most promising speculations about accelerated travel in space and time involves the use of Black Holes.

Black holes may have their uses. What we know about them till now is entirely theoretical, not tested against the skeptical standards of observation. There are some strange possibilities that have been suggested for black holes. Since there is no way to get out of a black hole, it is, in a sense, a separate universe.

In fact, our own universe is very likely itself a vast black hole. We have no knowledge of what lies outside our universe. This is true by definition, but also because of the properties of black holes. Objects that reside in them cannot ordinarily leave them. In a strange sense, our universe may be filled with objects that are not here. They are not separate universes. They do not have the mass of our universe. But in their separateness and their isolation they are autonomous universes.

There is an even more bizarre prospect. In one speculative view, an object that plunges down a rotating black hole may re-emerge elsewhere and elsewhen in another place and another time. Black holes may be apertures to distant galaxies and to remote epochs. They may be shortcuts through space and time. If such holes in the fabric of the space-time continuum exist, it is by no means certain that it would ever be possible for an extended object like a spacecraft to use a black hole for travel through space or time. The most serious obstacle would be the tidal force exerted by the black hole during approach – a force that would tend to pull any extended matter to pieces. And yet it seems to me that a very advanced civilization might cope with the tidal stresses of a black hole.

How many black holes are there in the sky? No one knows at present, but an estimate of one black hole for every hundred stars seems modest by at least some theoretical estimates. I can imagine, although it is the sheerest speculation, a federation of societies in the galaxy that have established a black hole rapid-transit system. A vehicle is rapidly routed through an interlaced network of black holes to the black hole nearest its destination.

At a typical place in the galaxy, one hundred stars are encompassed within a volume of radius of about twenty lightyears. If we imagine relativistic space vehicles for the short journeys – the local trains or shuttles it would take only a few years' ship time to get from the black hole to the farthest star of the hundred. One year on board the relativistic shuttle would be occupied accelerating at about 1 g, the acceleration we are familiar with because of the gravity of earth. After one year at 1 g, we would approach the speed of light. Another year would be spent doing a similar deceleration at 1 g at the end-point of the journey. A galaxy with such a transportation system, a million separately arisen civilizations and large numbers of worlds with colonies, exploratory parties, and work teams – a galaxy where the individuality of the constituent cultures is preserved but a common galactic heritage established and maintained; a galaxy in which the long travel times make trivial contact difficult and the black hole network makes important contact possible – *that* would be a galaxy of surpassing interest.

I can imagine, in such a galaxy, great civilizations growing up near the black holes, with the planets far from black holes designated as farm worlds, ecological preserves, vacations and resorts, specialty manufacturers, outposts for poets and musicians, and retreats for those who do not cherish big-city life. The discovery of such a galactic culture might happen at any moment – for example, by radio signals sent to the earth from civilizations on planets of other stars. Or such a discovery might not occur for many centuries, until a lone small vehicle from earth approaches a nearby black hole and there discovers the usual array of buoys to warn off improperly outfitted spacecraft, and encounters with the local immigration officers, among whose duties it is to explain the transportation conventions to newly arrived yokels from emerging civilizations.

The deaths of massive stars may provide the means for transcending the present boundaries of space and time, making all of the universe accessible to life, and – in the last deep sense – unifying the cosmos.

TECHNICAL ASPECTS OF PROPULSION

There has been some discussion and limited investigation into the use of nuclear powered rockets driven by the pressure of light itself. In the early 1960s, John R. Pierce. Executive Director of the Research-Communications Principles Division of the Bell Telephone Laboratories in the United States, and one of America's leading applied scientists, did a sober analysis on this method of propulsion. It is evident that to reach even the nearest sun-like stars within a human lifetime, it will require a peak speed close to that of the speed of light (approximately 186,000 miles per second). Pierce said that a rocket propelled by light could not carry enough fuel to reach the speed of light. He noted that in the near-vacuum of space the emissions from this type of rocket would have nothing to push against. (In an atom-smasher on earth, it is possible to accelerate particles to almost the speed of light because the equipment used for that purpose has the whole earth to push against.) Some physicists have suggested using the huge inter-galactic clouds of hydrogen and dust to push against, but at best it is doubtful that this particular means of propulsion would be selected. If it is, the duration of the trip will run into centuries.

Pierce also discussed, in an article for the publication "Proceedings of the Institute of Radio Engineers," the possibility of scooping up hydrogen en route as fuel for such a voyage. Throughout the galaxy there appears to be at least a sprinkling of hydrogen, and in some hydrogen clouds there are as many as 1,000 atoms per cubic centimeter (thimble-ful), Pierce discussed a system using a "generous" scoop 100 square meters (more than 100 square yards) in area to collect hydrogen en route. This collecting device, he added, would no doubt consist of force fields, such as magnetism, rather than a material substance. Taking a space ship 17.5 tons as the smallest conceivable size for interstellar travel, he estimated that the highest speed within reach of such a system would be 9.3% of the speed of light. "Clearly," he said in conclusion, "it is impossible to attain a velocity close to that of light by using interstellar matter as fuel."

Drs. I.S. Shklovskii and Carl Sagan in their book. *Intelligent life in the Universe*, considered the American physicist Robert W. Bussard's concept of an interstellar ramjet as a propulsion for intergalactic spacecraft:

> Bussard proposes an interstellar ramjet which uses the atoms of the interstellar medium, both as a working fluid (to provide reaction mass) and as an energy source (through thermonuclear fusion). With this system there is no complete conversion of matter into energy. Such a fusion reactor is certainly not available today (1966), but it violates no physical principles. Its construction is currently

being very actively pursued in research on controlled thermonuclear reactions, and there is no reason to expect it to be more than a century away from realization.

Such an interstellar ramjet would require a large surface area in order to draw in sufficient interstellar gas (or dust) to propel the craft. The calculations of Bussard indicate that if there were one atom of hydrogen per cm^3 in the interstellar medium, the surface density of the ramjet would have to be 10^8 gm or cm^{-2}. In general, the intake surface area of the ramjet is inversely proportional to the concentration n/h of the interstellar gas. If, for example, the mass of the rocket were 106 tons, and n/h equaled 1 atom cm^{-3}, the surface area of the ramjet intake would have to be 10^{15} cm^2, corresponding to a radius of about 700 hm. In meta galactic space, where n/h = 10^{-5} atoms cm^{-3}, the intake radius would have to be 100 times greater.

These frontal loading areas seem enormously large by contemporary standards, and perhaps remain absurdly large even when we project the progress of future technology. But we should emphasize that the collecting areas need not be material. Intense magnetic fields are now routinely generated in the laboratory, and even in commercial applications through the use of what we called superconducting flux pumps. Magnetic fields guide charged particles along a specified trajectory, and if the magnetic lines of force are cleverly arranged, through the design of the flux pumps, the charged particles can be conveyed to any desired region within the magnetic field. Thus, it seems at least possible that the collection of atoms of the interstellar medium by ramjet starships will be accomplished by ionizing the medium ahead of the spacecraft, and guiding the ions into the intake area through the use of intense magnetic fields.

The matter of propulsion has also been elaborated on by Edward Purcell, discoverer of the 21-centimeter line in the radio spectrum, in a lecture at Brookhaven National Laboratory on Long Island, the site of one of the world's most powerful accelerators (or "atom smashers"), to an audience consisting largely of physicists. Purcell's arguments were later elaborated in *Science* by Sebastian von Hoerner, a former associate of Frank Drake at Green Bank who later took a position at the Astronomisches Rechen-Instit (Astronomical Calculation Institute) in Heidelberg, Germany. Sullivan, in his book *We Are Not Alone*, expressed Purcell and von Hoerner's position in the following way:

> Each began by calling to mind the immensity of interstellar distances. If, for example, the sun were scaled down to the size of a cherrystone, the earth would be a grain of sand three feet away. The nearest star would be another cherrystone 140 miles away, but no advanced technologies would probably be found nearer than some 25,000 miles.

> Purcell pointed out that the strange "time dilation" predicted by the special theory of relativity would help passengers endure such long journeys if, in fact, they could be accelerated to a speed close to that of light. Let us assume, for example, that a vehicle accelerates at a rate equivalent to the force of gravity at the earth's surface (that is, at a rate of one "G"). We can be sure that human bodies could withstand such acceleration indefinitely, since the force exerted by gravity on our bodies throughout life is one G. Within a year such a vehicle would be moving almost as fast as light (186,000 miles a second). From then on, if the thrust of the engines remained constant, the rate of gain in speed would decrease, approaching zero acceleration but not reaching it, in the manner that mathematicians call "exponential." The engine thrust, no longer able to boost the speed to any degree,

would produce strange effects, from the viewpoint of an observer back home. The clock rate – the inherent property "time" – on board the vehicle would slow down, approaching zero but never quite getting there. The weight of the ship and its occupants would increase, and they would be foreshortened in the direction of flight.

At the midpoint of a journey to one of the nearest sunlike stars, 12 lightyears away, such a vehicle would be traveling 99% of light's speed. To land on a planet at its destination it would then have to decelerate at the same rate, with time still dragging its heels, until, during the final year of this outward journey, the speed dropped appreciably below that of light.

These peculiar manifestations would not be evident to those on board. Time would seem normal, as would their shape and weight. Assuming they then returned to earth in the same manner, the time for the entire journey, as measured on earth, would be twenty-eight years (a ray of light would have made the round trip in twenty-four years). However, the passengers would only be ten years older than when they started.

In a sense this is because of the peculiar relationship between time, as a dimension, and the speed of light. When we throw a light switch, the room seems illuminated instantaneously. In the world of our daily lives, the speed of light seems infinite and time seems invariable. But if one could accelerate almost to the speed of light, then strange things would happen to time and the other dimensions. Von Hoerner pointed out that *the longer such a journey was, the more extreme the effect, until finally, from the point of view of those back home, time on the space ship would virtually come to a halt.* Thus, in a vehicle whose engine accelerated and decelerated it, with a force equal to gravity, a round trip journey that seemed twenty years long to the voyagers, carrying them to a point 137 lightyears away, would bring them home to a world 270 years older than when they left it. *A sixty year journey, by clocks on the spacecraft, would bring them back 5,000,000 years after their departure. They would have reached out 2,500,000 lightyears – further than the nearest galaxy like our own.*

Purcell said flatly, in his Brookhaven lecture, that the Theory of Special Relativity, in predicting such strange effects, "is reliable." If it were not, he told his atom-smashing audience, "Some expensive machines around here would be in very deep trouble." Protons in the big Brookhaven accelerator are boosted to 99.948 percent of the speed of light and their mass increases in the manner predicted by relativity. Those designing the Brookhaven machine had to take this into account or it would not have worked.

Furthermore, the slowing of time, on the atomic level, is demonstrated by the extended lifetimes of the nuclear particles known as muons, when traveling almost at the speed of light. Muons are produced when atoms are shattered by high energy collisions, such as those in an atom-smasher or the impacts of the high energy atomic nuclei known as cosmic rays, plunging into the atmosphere from space. So brief is the lifetime of a muon that it decays into other particles almost instantaneously, its average lifetime being 2.2 millionths of a second. Since muons produced by cosmic rays are generated near the top of the atmosphere, they should not survive long enough to reach the surface of the earth despite their high speed. Yet, precisely because of this speed, their lifetimes are prolonged in the manner predicted by Einstein, and muons rain steadily on the earth. In other words, their time, as we see it, is extended. If this happens on the atomic level, would it apply to entire human beings. It is now widely believed that this would be the case. In fact, it is hard to see why it should be otherwise.

Purcell analyzed the amount of energy required for the previously described round trip to a star 12 lightyears away, in which a top velocity of 99% speed of light would be reached at the midpoint of both the outward and the return journeys. In terms of fuel weight, the most efficient source of energy within our grasp is the fusion reaction of the hydrogen bomb (where hydrogen isotopes such as tritium and deuterium combine to form helium). Even more efficient is the fusion that makes the sun shine. In this reaction, four hydrogen nuclei combine, under great heat and pressure, to form one helium nucleus. Because the nuclear "glue," or binding energy, required to hold the helium nucleus together is slightly less than that in the original hydrogen nuclei, something is left over after the reaction and emerges as free energy. Even though it is less than one percent of the original mass, the energy released is formidable because of the Einstein equation $E=MC^2$ (the energy equals the converted mass multiplied by the speed of light squared).

Purcell assumed that the energy of this solar fusion process could be used by a rocket with 100% efficiency. Disregarding the weight of the rocket itself, it would still require 16 billion tons of hydrogen fuel to accelerate the ten-ton capsule to 99% of the speed of light. To slow it down for the landing would require another 16 billion tons.

"But," said Purcell, "this is no place for timidity, so let us take the ultimate step and switch to perfect matter-anti-matter propellant." From our present knowledge of nature, there does not seem to be any more efficient way to obtain energy than to combine matter with antimatter. When two such substances meet, they mutually annihilate each other, leaving nothing but a great deal of energy in the form of gamma rays. It is the only known process in which matter (and antimatter) can be converted entirely into energy. The gamma rays, which are at the short-wave end of the electromagnetic spectrum, could power a rocket by the equivalent of light pressure.

No one has ever "seen" any antimatter. The experimenter can observe in his/her detector evidence that a particle of antimatter existed there briefly, and some accelerators can store bits of antimatter for limited periods, but sooner or later they meet particles of matter and vanish. If one lists all the particles that can be produced by smashing atoms, there is an "anti" particle corresponding to every one of them. The antiparticle is a mirror image of its counterpart. If the latter carries a negative charge, like the electron, then its antiparticle is positive. The antiparticle of the electron, for example, is known as the positron. It has the same mass as the electron and the strength of its electric charge is the same, but it is positive instead of negative. Discovered in 1932, it was the first hint there is an ephemeral world of antimatter.

Purcell found that his hypothetical journey would still require 400,000 tons of fuel, equally divided between matter and antimatter. Although research into this area of energy conversion has been limited, this appears to be the most probable method of propulsion on Terra II, if we are to achieve a near-light speed and gain the effect of "time dilation."

The initial mass of this ideal fuel (antimatter) would then have to be some 200,000 times greater than the mass of Terra II. Antimatter is uncommon on earth for a very good reason: when brought into physical contact with ordinary matter, both become annihilated in a violent conversion of mass into energy, primarily in the form of gamma rays.

Shklovskii and Sagan discuss the problems inherent in the use of this type of propulsion:

Terra II ... A Way Out

The containment of the antimatter – to say nothing of its production in the quantities required – is clearly a very serious problem. We would not want it to accidentally come into contact with the walls of the spacecraft, themselves composed of ordinary matter. A number of interesting ideas have been put forward which might lead to a successful circumvention of this difficulty. For example, perhaps a special type of non-material, magnetic bottle, employing an intense magnetic field, could be used. Such magnetic bottles are now being investigated in connection with experiments on controlled thermonuclear reactions,

– Intelligent Life the Universe

Since Shklovskii and Sagan made that statement the Russians have made tremendous strides in this area, and have evolved "magnetic bottles" that can contain antimatter in the fashion described.

Darol Froman, who had been technical associate director of the American Los Alamos Scientific Laboratory in New Mexico, even suggested the possibility of propelling the entire earth to another part of the galaxy.

Sullivan says of Froman: (transporting the earth to another part of the galaxy)

He was well equipped for the discussion – since the Los Alamos Laboratory operated for the Atomic Energy Commission by the University of California, is headquarters for American research on reactors for space flight. In a talk to the Division of Plasma Physics of the American Physical Society in November, 1961, he noted that the sun eventually will burn out and discussed whether or not, before that dark day, it might be possible to push the earth into another solar system, the energy for this grandiose scheme would be obtained by fusion reactions, using sea water as the fuel source.

Because the oceanic supply of deuterium, the heavy form of hydrogen used in the hydrogen bomb, is insufficient to push the earth great distances, Froman proposed that it would be more reasonable to use the reaction that occurs in the sun (combining four hydrogen nuclei to form a helium nucleus), even though we are a long way from learning how to do this. The process would make it possible to use hydrogen nuclei that are abundant in the oceans. He suggested that a quarter of this fuel be allocated to escaping the sun's gravity, another quarter be held to maneuver the planet into another solar system, and the remaining half be used for interstellar propulsion and for light and heat en route. The moon would be forfeited to obtain additional fuel, since, as he put it, the moon "will be no good to us anyway." In the absence of sunlight it would be virtually invisible.

Froman's earth-propulsion system could operate for as long as 8 billion years, he said, perhaps enabling a planet to outlive its parent sun and reach solar systems 1,300 lightyears away. It might even seem preferable, he said, to keep on traveling through the galaxy rather than go into orbit around some other sun. The oceans would then have to be replenished from time to time by gathering water from planets encountered en route. For most of us, the most comfortable space ship imaginable would be the earth itself. So if we don't like it here because the sun is dying or something, let's go elsewhere, earth and all. We will not have to worry about all the usual hardships of space travel. For example, the radiation problem will disappear because of the atmosphere and because we will be going at low speed. The ease and comfort of this mode of travel are shown in the next slide.

At this point, there flashed on the screen an idyllic scene showing lady golfers, pine trees and great open spaces.

Actually the problem of propelling large numbers of

people to another solar system was discussed as early as 1951 by Lyman Spitzer, head of the Princeton University Observatory. He spoke of a vehicle weighing 10,000 tons, powered by a uranium pile of perhaps 1,000 tons, generating 2 million horsepower of useful energy. "Such a ship," he wrote, "could carry thousands of people and vast supplies anywhere in the solar system, and could even navigate to other stars, though many generations would be born, grow up and die on shipboard before such a journey were complete. However, launching such a ship from the Earth's surface to a close circular orbit would be a tremendous undertaking. With the use of chemical fuels, such a launching would require a rocket of some million tons gross weight, an achievement that would seem far, far in the future."

Freeman J. Dyson, at the Institute for Advanced Study in Princeton, made a proposal as to how interstellar vehicles might pick up momentum en route. His scheme (to steal a little bit of energy with which two very dense stars circle each other) was outlined by Sullivan:

If, he said, a vehicle approached one such star as the star was coming toward it, the gravity of the star would whip the vehicle around in a tight orbit, sending it off into space again with far more energy than it had to begin with. It would be almost as though the vehicle had been hit by a gigantic baseball bat. The star, having transferred to the vehicle some of the energy with which it was circling its twin, would move a tiny bit closer to the star with which it was waltzing through space.

The most remarkable feature of this procedure was that, even though the vehicle underwent an explosive rate of acceleration – some 10,000 g's – no harm would come to the most delicate passenger or the most sensitive piece of equipment on board. This is because the accelerating force would be applied with almost complete uniformity to every particle of the body or instrument on board. It would not be any more uncomfortable than falling through space.

Thus, said Dyson, a vehicle could very rapidly be speeded up by more than 1,000 miles a second. The best star systems for giving vehicles such enormous accelerations, he said, would be pairs of white dwarfs, tiny "senile" stars whose density is so great that they may weigh as much as 3,000 tons per cubic inch. It may be imagined, he wrote, "that a highly developed technological species might use white-dwarf binaries scattered around the galaxy as relay stations for heavy long-distance freight transportation."

One of the most ardent champions of the feasibility of interstellar travel is Carl Sagan. His main argument was formulated while he was at the University of California in Berkeley and was presented to the American Rocket Society on November 15, 1962. He sought to show, not only that such travel is possible, but that, "Other civilizations, aeons more advanced than ours, must today be plying the spaces between the stars."

Sullivan relates Sagan's arguments:

He argued that radio waves are but a poor way to achieve a meeting of the minds between beings with utterly different histories and ways of thought. Furthermore, the radio does not permit contact between an advanced society and one that is intelligent but not yet in possession of radio technology. Nor does it allow the exchange of artifacts and biological specimens.

Interstellar space flight sweeps away these difficulties . . . It reopens the arena of action for civilizations where local exploration has been completed; it provides access beyond the planetary frontiers, where the opportunities are limitless.

Sagan claimed that the fuel problem could be overcome by scooping up hydrogen en route and using it to power on interstellar ramjet. The conventional ramjet is used for high-speed vehicles, such as the Bomarc missile, that scoop up air into a narrowing duct, thus compressing it. The air is used to burn the fuel, and leftover gases are ejected out the rear. A characteristic feature of the ramjet is that the faster it goes, the more efficient it becomes, since the speed tends to increase the pressure differential between the scooped up air and the surrounding atmosphere.

The interstellar ramjet was proposed in 1960 by R.W. Bussard, who, like Froman, was associated with the Los Alamos Scientific Laboratory. He explained that his engine would be a "rough analogy" of a conventional ramjet. It would scoop up interstellar material, which is almost entirely hydrogen, using appropriate portions of the hydrogen as fuel for a fusion reactor. The leftover matter would be squirted out the rear, thus overcoming the problem, raised by Pierce, of providing the engine with something to "Push against." Bussard cited the 21-centimeter radio observations by the Dutch at Leiden, showing, here and there in the galaxy, clouds of ionized hydrogen that could be collected magnetically. An intake area almost 80 miles in diameter would be needed, he said, to achieve the needed velocity for a space ship of about 1,000 tons. "This is very large by ordinary standards," said Bussard, "but then, on any account, interstellar travel is inherently a rather grand undertaking . . ."

The gas and dust nebula NGC 2237 in the constellation Monoceros. The dark splotches are thought to be great concentrations of absorbing dust.

Sagan conceded that the problem of scooping up enough hydrogen was staggering. Amplifying Bussard's calculations, he found that in ordinary interstellar space, with only one hydrogen atom per square centimeter, the sweeping system would have to be 2,500 miles in diameter. However, within hydrogen clouds, where the density may reach 1,000 particles per cubic centimeter, the intake could be, as Bussard pointed out, less than 80 miles wide. Perhaps, said Sagan, "starships" dart from one such cloud to another. Furthermore, he said, there may be some way to ionize material in the non-ionized clouds that predominate in space, so that it can be collected magnetically. Or the starship could select paths through clouds of material that are already ionized. By magnetic techniques it may also be possible to divert particles from the passenger area, thus overcoming the radiation hazard without heavy shielding. The huge vehicle would have to begin its journey, Sagan said, with the aid of fusion powered rocket stages. The ramjets would be used only when clear of the earth.

The effects of time dilation at relativistic speeds (velocities near that of light) are difficult to grasp in abstract terms, so let us imagine it in actual destinations and their related time lapses, both in shipboard time and relative earth time.

At an acceleration of 1 g, it takes only a few years (ship time) to reach the nearest stars; 21 years to reach the galactic center; and 28 years (again, ship time) to reach the nearest spiral galaxy beyond the Milky Way. With an acceleration of 2 or 3 g's these distances can be traversed in about half this time. Of course, this time lapse (or dilation) has no effect on the passage of time on earth. Thus, for a round trip with a several year stopover to the nearest stars, the elapsed time on earth would be a few decades; to Deneb, a few centuries;

to the Vela cloud complex, a few millennia; to the galactic center, a few tens of thousands of years; to M31 (the greatest galaxy in andromeda), a few million years; to the Virgo cluster of galaxies, a few tens of millions of years; and to the immensely distant Coma cluster of galaxies, a few hundreds of millions of years. This is the earth time lapse each of these enormous journeys could be performed within the lifetimes of a human crew, because of the effects of time dilation on board the ship. Thus a trip to the farthest reaches of the now-visible universe becomes more practical to the ship's crew (at relativistic speeds) than a trip to a nearby star at conventional speeds, which might take centuries, shipboard time. The prospects are awesome. We have taken those first tentative steps; fear must not prevent us from those leaps ahead of us.

"Brothers . . . stoop not to renounce the quest of what may in the sun's path be essayed, the world that never hath possessed." Ulysses, in Dante's Inferno, XXVI.

In his talk to the Rocket Society, Sagan said his argument was designed to "lend credence" to the possibility that interstellar vehicles may become feasible for us. We can expect, he added, "That if interstellar space flight is technically feasible – even though an exceedingly expensive and difficult undertaking, from our point of view – it will be developed."

He carried the possibilities of time-slowing even further than Purcell and von Hoerner. In continuous acceleration or deceleration (that is, if the engine were kept on), journeys that reached as far into the universe as galaxies millions of lightyears away would still be possible in the lifetime of the passengers, even though it was questionable whether any civilization would exist on their return. As noted earlier, from the earthly point of view, time on such an extended journey would virtually come to a halt. In fact, Eugen Sänger, head of the Institute of Jet Propulsion Physics in Stuttgart, Germany, had calculated that, with an acceleration no greater than that of the earth's gravity, even the most distant parts of the visible universe could be reached within forty-two years, spaceship time. The incentive, said Sagan, would be greater for journeys to nearby solar systems. Even then, those on the home planet would have to wait perhaps hundreds or thousands of years for the return of their astronauts. It must, therefore, be assumed that a highly advanced society would also be stable over very long periods, preserving the records of previous expeditions and waiting patiently for the return of others. According to this hypothesis, civilizations throughout the galaxy probably pool their results and avoid duplication. There may be "a central galactic information repository" where knowledge is assembled, making it far easier for those with access to such information to guess where, in the galaxy, newly intelligent life is about to appear – a problem very difficult for us, with only our own experience on one planet to go by.

Along with the problems of propulsion, another serious difficulty must be overcome before relativistic intergalactic spaceflight becomes practical. The space ship (be it powered by ramjet or antimatter conversions) is traveling at near the velocity of light. This is equal to the space ship sitting still and having particles (dust and interstellar grains of material) rushing into it at near the velocity of light. No amount of shielding would protect a craft from such bombardment.

For those who care about the figures, Shklovskii and Sagan describe the situation as follows: "the maximum velocity of the ramjet would be

$$V = C[1-(1 + aS,2c^2)-2] \; 1/2$$

where 'S' is the destination distance, and 'a' is the constant acceleration and deceleration chosen. If 'S' equaled 10,000 parsecs (1 parsec is 3.26 lightyears, or the distance from which the radius of the earth's orbit subtends an angle of one second arc) the distance to the galactic center, 'V,' would differ from 'c' by only one millionth of a percent. At this velocity, each atom of the interstellar medium colliding with the ramjet would appear as a component of the cosmic rays having an energy of 10^{13} electron volts cm^{-3}. Since the spacecraft is moving almost with the velocity of light, the flow of equivalent cosmic radiation striking the frontal loading area of the ramjet would be 10^{13} electron volts cm^{-3} x (3 x 10^{10} cm 23 -10 cm sec^{-1}) = 3 x 10^{23} electron volts cm^{-2} sec$^{-1,}$ or 2 x 10" erg cm^{-2} sec^{-1}. This is penetrating radiation, with an intensity 100,000 times greater than the intensity of sunlight at the surface of the earth. In other words, the crew would be fried."

It should be very evident both from the huge mass ratios required for fuel (be it antimatter or the ratios required for frontal load areas on a ramjet) that material shielding will never be a practical solution. Here we should again examine the method of "magnetic deflection" technique proposed earlier to guide interstellar particles to the ramjet's thermonuclear reactor. These same magnetic fields could be used to "sweep" or deflect particles from the path of the spaceship.

Sagan likened the "spacefaring" societies to those of the European Renaissance that sent voyagers eagerly in search of new worlds on our own planet. He suggested that such societies might send out expeditions about once a year and, hence, the starships would return at about the same rate, some with negative reports on solar systems visited, some with fresh news from some well-known civilization. "The wealth, diversity and brilliance of this commerce," he said with an exuberance reminiscent of Tsiolkovsky, "The exchange of goods and information, of arguments and artifacts, of concepts and conflicts, must continuously sharpen the curiosity and enhance the vitality of the participating societies."

On the assumption that there are about a million worlds in the galaxy capable of such feats, Sagan proposed that they would visit one another about once in every thousand years and that scouts may have visited the earth from time to time in the past perhaps a total of 10,000 times over the full span of the earth's history. One or two million years ago such visitors would have observed the emergence of primates ancestral to man and may have decided to step up the frequency of their visits to once every thousand years.

Which of these methods of propulsion is selected for Terra II will be up to the physicists on the "Propulsion and Energy Services" plexus. For obvious reason (time dilation) a propulsion capable of attaining near-light speed will be preferable.

While considering the possible types of propulsion, we should have in mind the probable evolutionary concepts likely to emerge when the world's leading scientists join forces in research on this project. Although a specific propulsion system might be assigned a statistical probability (and research priority) by the "Plexus," it is probable that a heretofore unknown means of propelling Terra II will be developed. Our technological evolution has taught us that concerted research fosters unimaginable benefits. If our prime objective is a near-the-speed-of-light propulsion, retrospection on the human innovative genius leaves little doubt of the probability of success.

Terra II ... A Way Out

21. Offer an Alternative.

(This is the English translation of the 21st Transmission)

Historical Imperative

The serious and practical intention of the Starseed company to launch a frail and pitifully slow craft into the limitless expanse of space in the hopes of contacting technologically superior civilizations may seem futilely optimistic. But in another sense we are simply following the historical imperatives provided for us by our immediate past. Imagine the year is 1490 and that we belong to an Amerindian tribe living on the Atlantic coast. We outfit our largest canoe, filled with water and supplies and head *east* hoping to find what actually there existed, the superior technologies of England, Spain and Portugal. What a dramatic change it would have effected in the psychological history of the world if Europe had been discovered by the Indians, sailing into Cadiz harbor with the proud, confident dignity of a race who dared to dream. That, after all, is the main difference between generations and races of humanity. There are those who discover and those who wait to be discovered.

Stealing the World

Two prisoners are walking the yard discussing the implications of gravity manipulation.

"We now know enough to create force-fields which would make it necessary to build a ship. We can throw a magnetic force-field around any section of the earth, an island, for example, and propel it through interstellar space."

"We could take the moon."

"Easily."

"But the people left on earth wouldn't like it. It would upset the tides and the biological clocks."

"We could take the earth and the moon. They'd never know."

"For all we know, the universe may have been moved out of orbit."

"Is that against the law?"

"What?"

"Stealing the universe."

STRANGER IN A STAR LAND

Heinlein: I think it's likely that we'll send out a manned star ship within the next generation. And the people who go out in it will not necessarily expect to return, any more than those in the Mayflower expected to return.

I'm of the opinion that life is ubiquitous. To a certain extent, this involves a philosophic or even a religious opinion, but I have a great trouble in conceiving of the universe – many thousands of light-years in size and containing so many millions of stars – as a place with just one tiny speck of life in it. I could be wrong, but it seems to me that if life is a rarity then god was being awfully wasteful of real estate to create so much back yard for so few tenants. I personally think life as we know it will be found throughout the universe under any conditions that permit the formation of large molecules.

Terra II ... A Way Out

Oui: Did you feel that the lunar landing vindicated your faith in space travel?

Heinlein: Yes, I'd waited 50 years for it, literally 50 years. Half a century of being treated like a madman for believing what has been perfectly evident since the days of Sir Isaac Newton.

At the present time we are a small, backwater planet – circling a second-rate star in what is actually a relatively uninhabited corner of the galaxy. We're out in the boondocks; we're not anywhere near the center of stellar population in this galaxy by any means. As for the earth being forgotten as man's home, Dr. Asimov and several others have written stories along this very line. But there's also another possibility; that we may *already* have forgotten our original home. We do have evidence of sorts that the human race started on this planet, but there is other evidence that has been ignored to the effect that we didn't start here at all . . .

You might also consider the fact that in many branches of the human race, in myths and religious stories, we came down out of the sky somehow. This idea runs through many religions. Even the Christian religion makes a stab at it; Adam and Eve certainly weren't supposed to have evolved here; they were supposed to have been exiled from the Garden of Eden – wherever that is – and *placed* here. Religious stories of that sort are quite common in our race, of course. Scientists ordinarily ignore them.

Time is running out . . . We can go to the stars. Not all of us, of course. There is not now, nor is there any prospect in the predictable future, any way of moving even our daily increase in population off this planet. But it *is* possible to move some thousands, or even millions, and establish colonies elsewhere – on the moon, on Mars, on Venus, even on planets around other stars, and thereby ensure the longtime future of our race. Out of this could come something else for all of mankind – hope!

Interview with Robert Heinlein, *Oui* magazine

EXPAND YOUR INTELLIGENCE

We are on the verge of the greatest neurological movement in history. A new grouping based on intelligence and designed to expand intelligence. Terra II.

The pleasantness of the "movement" is that it does not seek to proselytize or recruit members. The process is self-selective.

The more intelligent you are the more you realize that the only important thing to do in life is to expand your intelligence.

The molecular assemblage of such persons is obviously an intelligence enhancing step.

The intelligent thing to do is to get into direct communication with other intelligent people and link energies to attain immortality, and to contact the sources of higher intelligence.

Everyone involved is aware of the psychological and social traps. The ability to avoid the standard chauvinisms and egocentricities is a basic criterion of neural ability.

It is obvious that the organization of intelligent entities should be modeled upon the laws and patterns of atomic structuring. Group mind. Inter-neural network.

It seems inevitable that the future of evolution is going to lead us to transcend the limits of the body. As Robert Heinlein has pointed out, "a number of speculative thinkers – Stapledon, Dr. Jack Williamson, Professor Fred Hoyle, and others – have gone so far as to assume that even stars and galaxies might be living entities."

This definition of Higher Intelligence is far-fetched in every sense of the word, but certainly no more speculative than any other theory about universal creation.

The most sensible way to organize intelligent persons for intelligent purposes is to imitate atomic and stellar structure, which are themselves quite similar. We must act like "stars" and begin establishing interstellar communication among us.

AN OFFER

We evolve, dancing and laughing in the metamorphosis of the love that we now feel, each for the other. The planet earth is a miniscule part of our immediate play ground. Our civilizations' sages have been telling us that the linkage could only be made through exchange of linear imprints. Terra II suggests that we have the entire galaxy and maybe more, if we accept to expand our traveling perspective, to create the synaptic connection.

It is total arrest to indulge in the imprint of death.

Death is a tired metaphor. Our only mission is to make love to life itself. The acceptance of death is an unnecessary bondage, refutation of further mutation.

The only question is how can I contribute to the next act of evolution? How can I be an *agent evolutionaire*?

This thought reached the mind of Antoine de Saint-Exupery one night when his partner in evolution was transminding the evolutionary process:

> "Tonight, it will be a year . . . My star, then, can be found right above the place where I came to the earth, a year ago . . .

> "All men have the stars . . .But they are not the same things for different people. For some, who are travelers, the stars are guides. For others they are not

more than little lights in the sky. For others, who are scholars, they are problems. For my businessman they were wealth. But all these stars are silent. You – you alone – will have the stars as no one else has them"

"What are you trying to say?"

"In one of the stars I shall be living. In one of them I shall be laughing. And so it will be as if all the stars were laughing, when you look at the sky at night . . . You – only you – will have stars that can laugh."

– *The Little Prince*

It is weakness to accept the blindfold of time. Become the turning hands within the watch. Come in for alternative time.

The mind is the receiver: from the vast universe the message comes, transferred mutably into human frequencies. The decoding secret is the DNA molecule's basic chemicals: Man-Woman, Negative-Positive. Atomic particles synthesizing loving information, Pro-Creator, all ways similar. "Join us, come to us, intertwine us with your perfect electronic light years, fuse the energy!"

We expect an offer. Remember, it is urgent for all of us to *create* a plausible future for the evolution of our galaxy. Otherwise we will come to a stop in the adventure of mankind.

We have one aim: to all ways communicate with Higher Intelligences and listen avidly until our input outputs us to the next decoding exploration.

Creative loving is making evolution in the back seat of a 1999 time ship!

• • • BEGIN • • •

CENTRAL CONTROL

REPRODUCED COMMUNITY

TERRA VALLEY

VIEW OF TERRA VA

REFRIGERATION

ELEVATOR

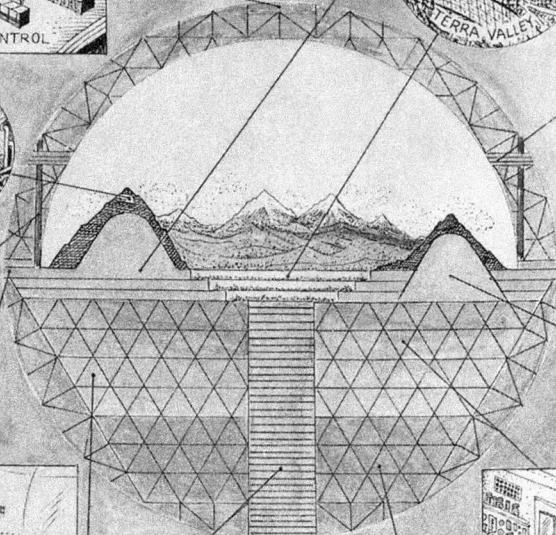

LOWER LEVEL CITY

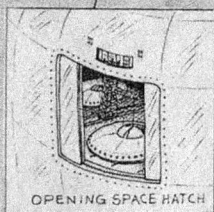
OPENING SPACE HATCH

WAREHOUSING

RAW MATERIAL

OFFICES & LABORATORIES

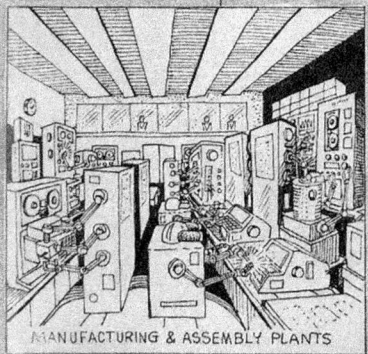
MANUFACTURING & ASSEMBLY PLANTS

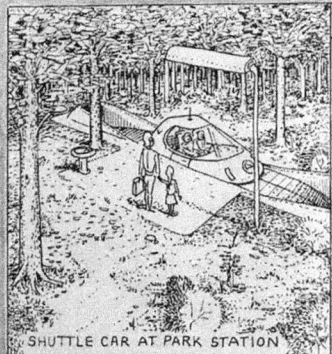
SHUTTLE CAR AT PARK STATION

OPEN AIR THEATRE

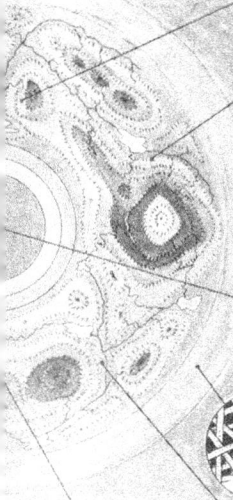

RING APARTMENTS

WALKS

SKI LODGE

SCIENCE LIBRARY

VALLEY FLOOR

UTILITY

CAL GARDENS

HOSPITAL

TERRA II

TRANSMITTED BY

TIMOTHY LEARY

L. WAYNE BENNER

GUANINE

JOANNA LEARY

HILARITAS
PRESS

Publishing the Books of Robert Anton Wilson
and Other Adventurous Thinkers

www.hilaritaspress.com